ELECTRONICS FOR INDUSTRIAL ELECTRICIANS

Second Edition

Stephen L. Herman

DELMAR PUBLISHERS INC.®

NOTICE TO THE READER

Publisher does not warrant or guarantee any of the products described herein or perform any independent analysis in connection with any of the product information contained herein. Publisher does not assume, and expressly disclaims, any obligation to obtain and include information other than that provided to it by the manufacturer.

The reader is expressly warned to consider and adopt all safety precautions that might be indicated by the activities described herein and to avoid all potential hazards. By following the instructions contained herein, the reader willingly assumes all risks in connection with such instructions.

The publisher makes no representations or warranties of any kind, including but not limited to, the warranties of fitness for particular purpose or merchantability, nor are any such representations implied with respect to the material set forth herein, and the publisher takes no responsibility with respect to such material. The publisher shall not be liable for any special, consequential or exemplary damages resulting, in whole or in part, from the readers' use of, or reliance upon, this material.

Cover Credit: General Electric FANUC Automation

Delmar Staff

    Senior Executive Editor: David C. Gordon
    Project Editor: Ruth East
    Production Coordinator: Larry Main
    Design Coordinator: Susan Mathews

For information, address Delmar Publishers Inc.
2 Computer Drive West, Box 15-015
Albany, New York 12212

Printed in the United States of America
Published simultaneously in Canada
by Nelson Canada,
a division of The Thomson Corporation

Library of Congress Cataloging in Publication Data

Herman, Stephen L.
  Electronics for industrial electricians.

  Includes index.
  1. Industrial electronics I. Title.
TK7881.H47 1989   621.381   89-11607
ISBN 0-8273-3808-2
ISBN 0-8273-3809-0 (instructor's guide)

Contents

Circuit Applications

Preface

Electronics for Industrial Electricians is a practical hands-on course in industrial electronics. The text was written with the assumption that the student has a knowledge of basic electricity, Ohm's Law, series, parallel, and combination circuits. Components and circuits are explained in a straightforward and practical manner, as opposed to the traditional mathematical approach used by most textbooks. Each unit contains stated objectives at the beginning.

Many of the units contain an *application* example that uses the component or components covered in that unit in a practical circuit. The purpose and function of each component is explained to aid students in understanding how the circuit actually works.

The last four units are an opportunity to combine all of the knowledge and skills gained in the earlier units. These four circuits combine the devices and design characteristics discussed earlier into functioning circuits.

CURRENT FLOW THEORY VERSUS ELECTRON FLOW THEORY

The student of electricity and electronics should be aware that there are actually two theories concerning the direction of flow of electricity. One is the current flow theory, which states that electricity flows from a positive source to a negative source; and the other is the electron flow theory which states that current flows from the negative source to the positive source. Although the electron flow theory is the most widely accepted, most electronic schematics are drawn in such a manner that it is easier to understand the logic of the circuit if the current flow theory is applied. The reason for this is that it has been the convention for many years to use the negative side of the power supply as ground. The positive side of the power supply is considered to be some value of voltage above ground. A good example of this type of circuit can be seen in most automobiles. The negative post of the battery is connected to the frame of the vehicle and is considered to be ground. The positive post of the battery is generally considered to be the "HOT" terminal.

Electrical and electronic schematics are drawn in such a manner as to be read like a book; that is from top to bottom and from left to right. Since most schematic diagrams are drawn in this manner, it is the contention of the author of this text that students of electricity and electronics should become familiar with the conventional current flow theory. For this reason, this text has been written with the assumption that current flows from the positive source of voltage to the negative source.

ABOUT THE AUTHOR

Stephen L. Herman has been both a teacher of industrial electricity and an industrial electrician for many years. His formal training was obtained at Stephen F. Austin University in Nacogdoches, Texas and Catawba Valley Technical College in Hickory, North Carolina. Mr. Herman has worked as a maintenance electrician for Superior Cable Corp. and as a class "A" electrician for National Liberty Pipe and Tube Co. During these years of industrial experience he has gained both a practical and theoretical knowledge of electronics circuits as they apply to industrial applications. He was employed as the instructor in electrical maintenance at Randolph Technical College in Asheboro, North Carolina for nine years. As of the date of this text, Mr. Herman is employed as an instructor in industrial electricity at Lee College in Baytown, Texas.

ACKNOWLEDGMENTS

The author would like to express appreciation to the following people:

The administration, faculty, and staff of Lee College in Baytown, Texas for their technical assistance and encouragement.

Mr. Eugene B. Hicks, electronics instructor at Randolph Technical College in Asheboro, North Carolina. Gene is a dear friend who taught me much of the information contained in this text.

The individuals who reviewed this text: Robert W. D'Vileskis of Providence, Rhode Island; Joseph Szumewski of Cinnaminson, New Jersey; Don O. Smith of Fitchburg, Massachusetts; and R. P. O'Riley of Dallas, Texas.

The following companies who furnished photographs and/or technical information contained in this text:

AAVID Engineering Co.	Struthers-Dunn Inc.
General Electric Co.	Tektronix Inc.
International Rectifier	Texas Instruments
Magnecraft Electric Co.	Vactec Inc.
Ramsey Controls Inc.	

UNIT 1
The Oscilloscope

For many years the industrial electrician's measuring tools have been the volt-ohm-milliammeter (VOM), and the clamp-on ammeter. These old standbys are still good tools and are the best way to troubleshoot many of the circuits the industrial electrician encounters. However, many of the electronic control systems in today's industry produce voltage pulses that are meaningless to a VOM. In many instances, it is necessary not only to know the amount of voltage present at a particular point in a circuit, but also the length or duration of the pulse and its frequency. Some pulses may be less than one volt and last for only a millisecond. A VOM does not measure many of these things. Therefore, the oscilloscope must be used to learn what is actually happening in a circuit.

OBJECTIVES

After studying this unit the student should be able to:
- Discuss the operation of an oscilloscope
- Discuss various oscilloscope controls
- Connect an oscilloscope into a circuit
- Interpret wave forms produced on the display of the oscilloscope

OSCILLOSCOPE BASICS

This unit is designed to teach some of the fundamentals of using an oscilloscope. It is not intended to make you an expert in its use. The first point to understand about the oscilloscope is that it is a voltmeter; it measures voltage. It does not measure current, resistance, or watts. The oscilloscope not only measures a voltage during a particular period of time, it creates a two-dimensional image, or picture on its screen.

Some Important Parts of the Oscilloscope

The oscilloscope is divided into two main sections. One section is the voltage section and the other is the time base. The display of the oscilloscope is divided by vertical and horizontal lines, figure 1-1. Voltage is measured on the vertical or Y axis of the display and time is measured on the horizontal or X axis of the display.

When using the VOM, a range switch permits the selection of a different range of voltages which will deflect the meter full scale, 600 volts, 300 volts, or 60 volts, for instance. Changing voltage ranges permits much more accurate measurements of voltage. Trying to measure 24 volts on the 600 volt range will not move the meter enough to make any kind of accurate measurement. By changing to a range of 60 volts full scale, however, 24 volts can be read very accurately.

The oscilloscope also has a voltage range switch, figure 1-2. The

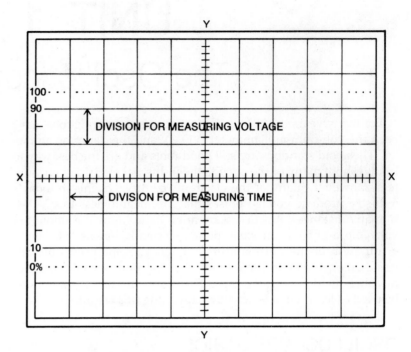

Figure 1-1 Display of oscilloscope

voltage range switch on an oscilloscope selects volts per division instead of volts full scale. For instance, the voltage range switch shown in figure 1-2 is set for 10 millivolts (mv) at the 1X position. This means that each of the lines in the vertical direction (the Y axis) of the display, figure 1-1, has a value of 10 millivolts per division. Assume the oscilloscope has been adjusted to 0 volt on the centerline of the display. This adjustment is made by grounding the probe of the oscilloscope and setting the trace on the centerline of the display. The probe is then removed from ground and connected to the circuit to be tested. If the trace rises above the centerline, the voltage is positive with respect to ground. If the trace drops below the centerline, the voltage is negative with respect to ground. If the oscilloscope probe is connected to a positive voltage of 30 millivolts, the trace rises to the position marked (A) in figure 1-3.

If the probe is connected to a negative 30 millivolts, the trace falls to the position marked (B) in figure 1-3. Notice that the oscilloscope can display a negative voltage as easily as it can display a positive voltage. If the range is changed to 20 volts per division, (A) in figure 1-3 will display a value of 60 volts positive.

The next part of the oscilloscope to become familiar with is the

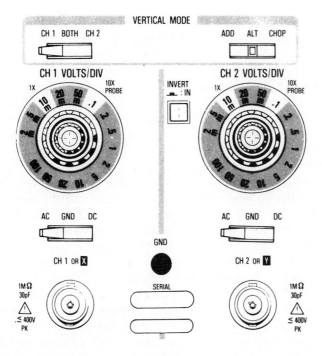

Figure 1-2 Voltage range control

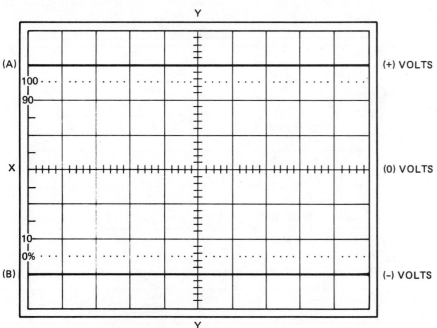

Figure 1-3 Display showing positive and negative dc voltages

Figure 1-4 Time base

time base, figure 1-4. The time base is calibrated in seconds per division and has range values from seconds to microseconds. The time base controls the value of the divisions of the lines in the horizontal direction. For instance, if the time base is set for 5 milliseconds per division, the trace will sweep from one division to the next in 5 milliseconds. With the time base set in this position, it will take the trace 50 milliseconds (ms) to sweep from one side of the display screen to the other. If the time base is set for 2 microseconds per division, the trace will sweep the screen in 20 microseconds.

The oscilloscope not only measures the value of voltage, it also measures the voltage with respect to time. Since the oscilloscope can measure the time to complete one cycle of ac voltage, its frequency can be computed. This is done by dividing one by the time it takes to complete one cycle, $F = 1/t$. Assume that the time base is set for .5 milliseconds per division and the voltage range is set for 20 volts per division. If the oscilloscope has been set so that the centerline of the display is zero volt, the ac waveform shown in figure 1-5 will have a peak value of 60 volts. Notice that the oscilloscope displays the peak value of voltage and not the root-mean-square (rms) or effective value which is measured by an ac voltmeter. To measure the frequency, count the time it took to

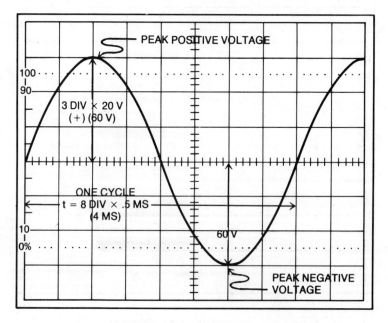

Figure 1-5 Display showing ac sine wave

complete one full cycle. The waveforms shown in figure 1-5 took 4 ms. Therefore, the frequency is 1/.004 s = 250 (hertz) Hz.

Most oscilloscopes use a probe which acts as an attenuator. An *attenuator* is a device that divides or makes smaller the input signal. An attenuated probe is used to permit higher voltage *readings* than are normally possible. Most attenuated probes are 10:1. This means that if the voltage range switch is set for 5 volts per division, the display is read as 50 volts per division. If the voltage range switch is set on 2 volts per division, each division on the display has a value of 20 volts per division.

Probe attenuators are made in various styles by manufacturers. Some probes have the attenuator located in the head while others have it located at the scope input. Regardless of the type of attenuator used, it may have to be compensated or adjusted. In fact, probe compensation should be checked frequently. The manufacturers also use different methods for compensating (adjusting) their probes, so follow the procedures given in the operator's manual for the oscilloscope being used.

Figure 1-6 Front of Tektronix Model 2213 oscilloscope

Some Common Controls

Become familiar with the more common controls found on an oscilloscope, figure 1-6.

- **Power:** The power switch is used to turn the oscilloscope on or off (1).
- **Beam finder:** This control is used to locate the position of the trace if it is off the display. The beam finder button indicates the approximate location of the trace and will help in moving the position controls to get the trace back on the screen (2).
- **Probe adjust:** This is a reference voltage point used to compensate the probe. Most probe adjust points produce a square wave signal at about .5 volt (3).
- **Intensity and focus:** The intensity control adjusts the brightness of the trace.

Caution: Be careful not to leave a bright spot on the display because it will burn a spot on the face of the cathode-ray tube (CRT). This burned spot results in permanent damage to the CRT.

The focus control sharpens the image of the trace (4).

- **Vertical position:** Adjust the trace up or down on the display. If a dual trace scope (an oscilloscope that can display two separate voltages at the same time) is being used, there are two vertical position controls (5).
- **Ch 1—Both—Ch 2:** This control permits the selection of either channel 1, channel 2, or will display both at the same time (6).
- **Add—Alt—Chop:** This control is active only when both traces are being displayed at the same time. The add mode adds the two waves together. The alt mode alternates the sweeps between channel 1 and channel 2. The chop mode alternates several times during one sweep. This makes the display appear to be more stable. The chop mode is used more frequently when observing two traces at the same time (7).
- **ac—GND—dc:** The alternating current (ac) mode is used to block a direct current (dc) voltage when only the ac part of the voltage is to be seen. Assume an ac voltage of a few millivolts is riding on a dc voltage of a hundred volts. If the voltage range is set high enough to permit 100 volts of direct current to be seen on the display, the ac voltage will not be visible. The ac section of this switch inserts a capacitor in series with the probe. The capacitor blocks the dc voltage and permits the ac voltage to pass. Since the 100 dc volts has been

blocked, the voltage range can be adjusted for millivolts per division and the small ac signal can be seen. The GND (ground) section of the switch grounds the input so the sweep can be adjusted for 0 volt at any position on the display. The ground switch provides ground at the scope but does not ground the probe. This means that the ground switch can be connected to a live circuit without a problem. The dc section permits observation of the voltage to which the probe is connected (8).

- **Horizontal position:** This control adjusts the position of the trace from left to right (9).
- **Auto-normal:** The auto-normal control determines whether the time base will be triggered automatically or if it is to be operated in a free-running mode. If it is operated in the normal setting, the trigger signal is taken from the line to which the probe is connected. The scope is generally operated with the trigger set in the automatic position (10).
- **Level:** This control determines the amplitude that the signal must be before the scope triggers (11).
- **Slope:** The slope permits the scope to trigger on the positive or negative half of the waveform (12).
- **Int—Line—Ext:** The scope is generally operated in the internal mode. In this mode, the trigger signal is provided by the scope. In the line mode, the trigger signal is provided from a sample of the line. The Ext (external) mode permits an external trigger signal to be applied (13).

These are not all the controls shown on the oscilloscope in figure 1-6, but they are the major ones. Most oscilloscopes contain these controls.

Figure 1-7 Camera used to photograph waveforms on the display of an oscilloscope

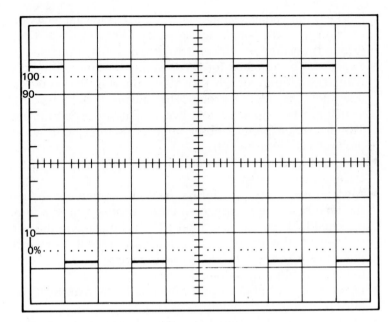

Figure 1-8 Display showing ac square wave

INTERPRETING WAVEFORMS

After learning to operate the controls on the scope, interpreting the waveforms shown on the display is learned. It generally takes experience and practice to become proficient in the use of the oscilloscope. When using the oscilloscope, remember that the display is the product of voltage and time. A camera such as the one shown in figure 1-7 can be used to photograph the various waveforms. These photographs can then be used for comparison of the waveforms as seen on the oscilloscope.

In figure 1-8, assume that the voltage range has been set for .5 volt per division, the time base is set at 2 ms per division, and 0 volt has been set on the centerline of the display. The waveform shown is a square wave. The display shows that the voltage rises in the positive direction to a value of 1.4 volts and remains there for 2 milliseconds. It then drops to 1.4 volts negative and remains there for 2 milliseconds before going back to positive. Since the voltage changes between positive and negative, it is an ac voltage. The length of one cycle is 4 ms, so the frequency is $1/.004 = 250$ Hz.

In figure 1-9 the oscilloscope has been set for 50 millivolts (mv) per division and 20 microseconds (μs) per division. The display shows a voltage which is negative to the ground lead of the probe, and has a peak value of 150 millivolts (mv). The pulse waveform crosses zero every 20 microseconds, therefore the pulse rate is

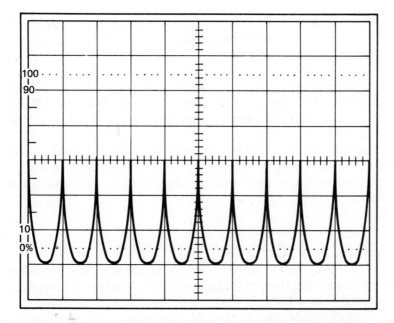

Figure 1-9 Display showing dc voltage in the negative direction

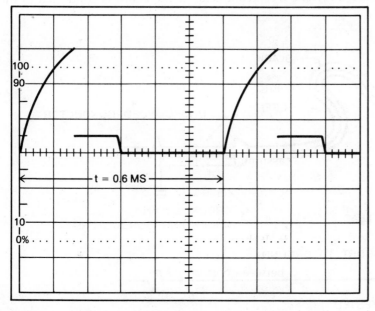

Figure 1-10 Display showing a dc waveform

$1/.000020 = 50$ kHz (kilohertz). The voltage shown is dc since it never crosses the zero reference and goes in the positive direction.

In figure 1-10, assume that the scope is set at 50 volts per division and .1 ms per division. The pulse waveform rises to a value of 150 volts in the positive direction and then drops to about 25 volts. It remains at 25 volts for .15 milliseconds before dropping back to 0 volt. It remains at 0 volt for .3 milliseconds before the waveform starts over again. The voltage shown is dc because it remains in the positive direction. To find the pulse rate, measure from the beginning of one pulse to the beginning of the next pulse. This is the period of one complete pulse. The length of one pulse period is .6 milliseconds. Therefore, the pulse rate is 1/.0006 = 1666 Hz.

Learning to interpret the waveforms seen on the display takes time and practice. It is worth the effort, however, because it is the only way to understand what is happening in many electronic circuits.

OSCILLOSCOPE PRECAUTIONS

When using an oscilloscope, certain precautions must be taken when connecting the probe to a live circuit. Most oscilloscopes are powered from a 120 volt ac source. Connection to the power line is generally made with a standard three-prong plug as shown in figure 1-11. The black conductor of the chord is connected to the

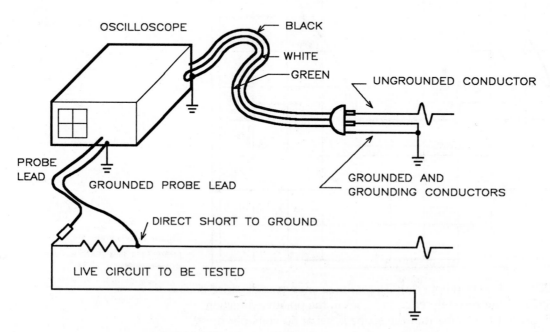

Figure 1-11 The grounded probe of the oscilloscope is connected to the ungrounded circuit conductor.

ungrounded side of the circuit and the white conductor is connected to ground and is the neutral conductor. The green conductor is connected to the metal case of the oscilloscope and is used
as the safety grounding conductor.

The probe end of the oscilloscope contains two conductors. One
is connected to the input and in most cases has an impedance of a
million ohms or more. The other conductor, however, is connected
directly to ground. If the oscilloscope should be used to test a
circuit that has one side grounded and the other side ungrounded,
care must be taken to ensure that the grounded conductor of
the probe is connected to the grounded side of the circuit. If the
grounded conductor of the probe should be connected to the
ungrounded side of the circuit, a direct short to ground through
the probe lead and case of the oscilloscope will exist.

UNIT 2
Semiconductors

OBJECTIVES

After studying this unit the student should be able to:

- Discuss the differences between the atomic structure of conductors, insulators, and semiconductors
- Give an explanation of how p- and n-type materials are made
- Describe a lattice structure

Solid-state devices are often called semiconductors because all solid-state devices are made from semiconductor materials. The atomic structure of materials must first be discussed to understand what a semiconductor material is. The materials of chief concern in the electronics industry are conductors, insulators, and semiconductors.

CONDUCTORS: These are materials that provide an easy path for electron flow. They are generally made from materials that have large, heavy atoms. This is the reason that most conductors are metals. The best of these conductors are silver, copper, and aluminum. Conductors are materials that have only one or two *valence* electrons in their atoms. These are the electrons in the outer orbit of an atom, figure 2-1. The best electrical conductor is an atom which has only one valence electron. This is because the electron is loosely held in orbit and is easily given up for current flow.

INSULATORS: These are generally made from materials that have small, lightweight atoms. An insulating material will have the outer orbit almost filled or filled with valence electrons. Remember that it takes eight valence electrons to fill the outer shell, figure 2-2. Since the outer orbit is almost filled or filled with valence electrons, the electrons are tightly held in orbit and not easily given up for current flow.

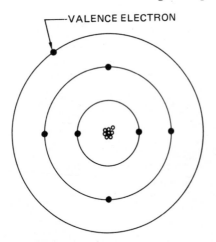

Figure 2-1 Atom of a conductor

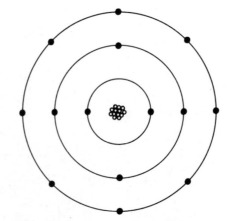

Figure 2-2 Atom of an insulator

SEMICONDUCTORS: These are materials that are neither good conductors nor good insulators. They are made from materials that have four valence electrons in their outer orbit, figure 2-3.

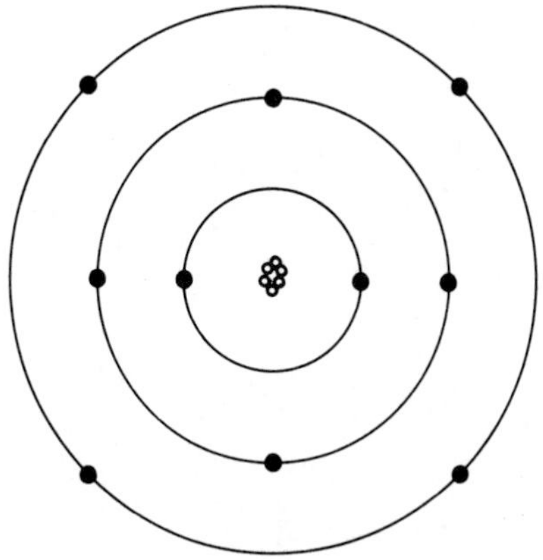

SEMICONDUCTOR MATERIAL

The most common semiconductor materials are germanium and silicon. Of these two, silicon is used more often because of its ability to withstand heat. When semiconductor materials are refined into a pure form, the molecules arrange themselves into a crystal structure which has a definite pattern, figure 2-4. This is known as a *lattice structure.*

A pure semiconductor material such as silicon is nothing more than a poor conductive material in its natural state. To make a semiconductor material useful for the production of solid-state components, it must be mixed with an impurity. This process is known as doping. When the impurity has only three valence electrons, such as indium or gallium, the new lattice structure is different, figure 2-5. A hole is left in the material when the lattice structure is formed. This hole is caused by the lack of an electron where one should be.

P- and N-type Material

Since the material now has a lack of electrons, which are negative particles, it is no longer electrically neutral. Since a hole is in

Figure 2-3 Atom of a semiconductor

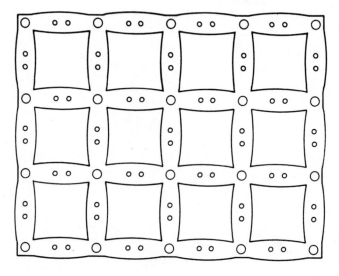

Figure 2-4 Lattice structure of a pure semiconductor material

a place where an electron should be, the hole has a positive charge. This semiconductor material now has a net positive charge, and is therefore known as a p-type material.

When a semiconductor material is mixed with an impurity which has five valence electrons, such as arsenic or antimony, the lattice structure will have an excess of electrons, figure 2-6. Since

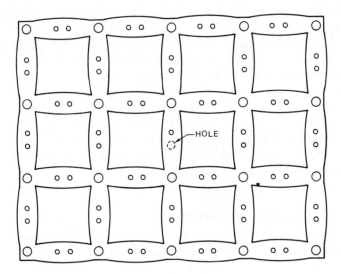

Figure 2-5 Lattice structure of a p-type material

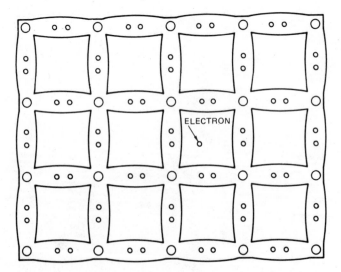

Figure 2-6 Lattice structure of an n-type material

electrons are negative particles, and there are more electrons in the material than there should be, the material has a net negative charge, and is referred to as an n-type material.

All solid-state devices are made from a combination of p- and n-type materials. The type of device formed is determined by how the p- and n-type materials are connected (joined) together. The number of layers of material and the thickness of various layers play an important part in determining what type of device will be formed. The diode is often called a p-n junction because it is made by joining together a piece of p-type and a piece of n-type material, figure 2-7. The transistor is made by joining three layers of semiconductor material, figure 2-8. Regardless of the type of solid-state device being used, it is made by joining together p- and n-type materials.

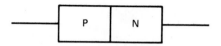

Figure 2-7 P-n junction

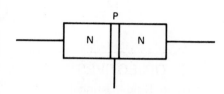

Figure 2-8 Transistor

UNIT 3
Power Rating and Heat Sinking Components

OBJECTIVES

After studying this unit the student should be able to:

- Discuss why heat sinks are necessary in electronic circuits
- Discuss the use of thermal compound
- Explain how thermal compound aids in the transfer of heat from the electronic component to the heat sink

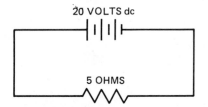

Figure 3-1 Simple dc circuit

$P = E \times I$, $P = 20 \times 4$, $P = 80$ WATTS

$P = I^2 \times R$, $P = 4 \times 4 \times 5$, $P = 80$ WATTS

$P = \dfrac{E^2}{R}$, $P = \dfrac{20 \times 20}{5}$, $P = \dfrac{400}{5}$, $P = 80$ WATTS

Figure 3-2 Formulas used to compute the power in a circuit

POWER RATING

Electrical and electronic components have a power rating measured in watts (w). This rating indicates the amount of heat that the component can dissipate before damage occurs. This can be found by using any one of three formulas, Power (P) = voltage (E) × current (I), Power = I² × resistance (R), Power = $\dfrac{E^2}{R}$.

Consider the resistor shown in figure 3-1.

In this example, a 5-ohm resistor is connected across 20 volts. A current of 4 amperes (A), will flow in this circuit. (I = $\dfrac{E}{R}$, I = $\dfrac{20}{5}$, I = 4). To find the amount of heat this resistor must dissipate, use any of the three formulas, figure 3-2.

When electrical components are connected into a circuit, care should be taken not to exceed their power ratings. If there is any doubt that the component can handle the heat it is expected to dissipate, the power should be computed by using one of the formulas shown in figure 3-2. Forcing a component to dissipate more heat than it was designed for can only shorten its life.

HEAT SINKS

Some electronic components must be *heat sinked* in order to operate at their listed power ratings. Assume a transistor has a power rating of 30 watts. This power rating can be obtained only if the transistor is mounted on a proper heat sink and thermal compound used.

Heat sinks vary in size and shape, but have only one purpose. That purpose is to increase the surface area of the device connected to it. This permits air to wipe a greater area and remove heat at a faster rate.

When a component is mounted to a heat sink, *thermal compound* is generally used to ensure a good thermal contact between the device and the heat sink. There are two types of thermal compound in common use. One type is a greasy substance that is usually white in color. This compound is an excellent conductor of heat, but is made from beryllium oxide which is a deadly poison.

Figure 3-3 Uneven surface of an electronic component

Figure 3-4 Two components with uneven surfaces joined together

When using this compound, care should be taken not to get it into the mouth or eyes. The second type in general use is a silicon grease which looks like a clear jelly. This compound is a good heat conductor and is less messy.

The surface of the device and heat sink may look perfectly flat to the naked eye. A closer examination, however, reveals that neither surface is flat. The surface appears similar to the surface shown in figure 3-3.

If these two surfaces are joined together, they may make actual metal-to-metal contact at relatively few points, figure 3-4. That is why thermal compound is used to fill in the gaps between the two surfaces and provide good thermal contact, figure 3-5. The importance of using thermal compound when heat sinking components cannot be overstressed.

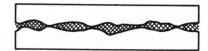

Figure 3-5 Heat sink compound used to fill the gaps

Some heat sinks are rather simple in design and are intended to be finger pressed onto a component, figure 3-6. This type of heat sink does not generally require the application of thermal compound. Small heat sinks are generally used when a component is operated at its full power rating. Other small heat sinks are clip-on types, as shown in figure 3-7. Regardless of whether a component requires the use of a heat sink be sure to keep the components operating within their power rating.

Figure 3-6 Heat sinks

Figure 3-7 Clip-on heat sink for T0220 case devices

PHOTO COURTESY OF AAVID ENGINEERING CO.

UNIT 4

The p-n Junction

OBJECTIVES

After studying this unit the student should be able to:

- Discuss the operation of a diode
- Explain forward and reverse bias
- Draw the schematic symbol of a diode
- Test a diode with an ohmmeter
- Construct a half-wave rectifier

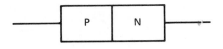

Figure 4-1 P-n junction

BASICS OF THE JUNCTION

The p-n junction is the simplest of all semiconductor devices. It is a two-element device made by joining a piece of p-type and a piece of n-type material, figure 4-1. The p-n junction is more commonly known as the *diode*, which is a device that operates like an electronic check valve. It permits current to flow through it in one direction only.

To understand the operation of the diode it will be necessary to return to the study of semiconductor material. Recall from Unit 2 that a p-type material has an excess of holes in its structure and an n-type material has an excess of electrons. When the p- and n-type materials are joined, electrons in the n-type material are repelled by other electrons and are attracted to the positive holes in the p-type material. Some of these electrons will drift across the junction and fill holes in the p-type region. When an electron leaves the n-type region a hole is left in the material. This hole produces an atom with a net positive charge or *positive ion*. Electrons that drift across the junction fall into holes and produce atoms with a net negative charge or *negative ions*. Each time an electron crosses the junction into the p-type region a pair of ions is formed.

These ions, unlike the free electrons or holes, are held in place by covalence bonding and are not free to move around inside the p- or n-type material. Since these ions exist at the junction of the p- and n-type material as shown in figure 4-2, they form a barrier to other electrons or holes to prevent them from crossing the junction. This potential produced by the ions is known as the *barrier potential* or *barrier charge*. Before holes or electrons can cross the junction, enough voltage must be applied to the diode to break

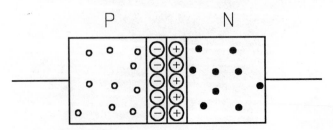

Figure 4-2 Barrier charge is formed by positive and negative ions

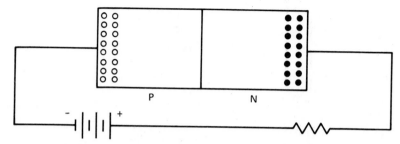

Figure 4-3 Reverse bias

down or overcome the barrier potential. At 25°C the barrier potential for a silicon device is generally between 0.6 and 0.7 volt. For a germanium device the barrier potential is generally between 0.3 and 0.4 volt.

OPERATION OF THE JUNCTION

Examine what happens when voltage is applied to the diode.

1. Assume a battery is connected, through a current-limiting resistor, to the diode. The negative pole of the battery is connected to the p-type material and the positive pole of the battery is connected to the n-type material, figure 4-3.
2. Since the negative pole of the battery is connected to the p-type material, the positive holes are attracted to the negative charge and gather at that side of the semiconductor material. The positive polarity of the battery is connected to the n-type material and attracts the negative electrons to that side of the material.
3. No current can flow through the diode when the battery is connected in this manner. When a diode has a negative polarity connected to its p section and a positive polarity connected to its n section, the diode is in a condition known as *reverse bias*.
4. Figure 4-4 shows the schematic symbol for the diode when the battery is connected to make the diode reverse biased.
5. If the battery is reconnected to the diode so that the polarity is reversed, the diode will be *forward biased* and current will flow.
6. In figure 4-5, the positive terminal of the battery has been connected to the p-type material. The negative terminal has been connected to the n-type material.
7. If the battery voltage is high enough to overcome the barrier potential of the p-n junction, the negative electrons are attracted to the positive side of the battery. They begin flowing across the junction. The positive holes are attracted to

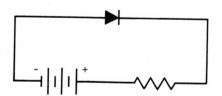

Figure 4-4 Schematic symbol shown reverse biased

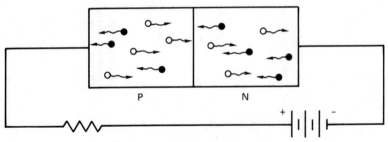

Figure 4-5 Forward bias

the negative terminal and begin flowing across the junction also.

When an electron reaches the positive battery terminal, it leaves the semiconductor material. There is now a hole where the electron had been. This hole moves toward the negative terminal of the battery. When it reaches the negative terminal, it is immediately filled with an electron. Therefore:

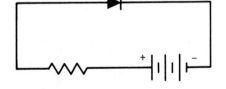

Figure 4-6 Schematic symbol shown forward biased

- An electron leaving the diode at the positive terminal creates a hole.
- A hole provides a place for another electron at the negative terminal.
- There are always as many holes and electrons left in the semiconductor material when the battery connection is broken as there were when the connection was made.

Figure 4-6 shows the schematic symbol for the diode with a battery connected to make the diode forward biased.

Diode Identification

Diodes are manufactured in a great variety of sizes, styles, and ratings. Some are referred to as rectifier diodes. These diodes are the common junction type and are designed to carry from moderate to heavy current, however, they are not designed to operate at high frequencies. The signal diode is designed to operate at a high frequency, but is generally a low power device.

Most diodes are given some type of reference number for identification. If reference material is available, the ratings of the diode can be learned. The most popular reference system for diodes is the 1N system. Diodes with 1N preceding their number are registered devices and can be identified without too much difficulty. A 1N4004 diode is made of silicon, has a voltage rating of 400 volts, a current rating of 1 ampere and is a rectifier diode. A 1N4020 diode is made of silicon, and is a zener diode (which will be discussed in unit 9). It is rated at 12 volts and will handle 5 watts. A 1N2326 diode is made of germanium, is rated at 1 volt and 2 milliamperes and is a signal diode.

Some of the common ratings of diodes that are of general interest are PRV and PIV, I_O, and V_F. PRV stands for peak reverse voltage, and PIV stands for peak inverse voltage. Both of these ratings list the amount of voltage a diode can withstand in the reverse bias direction without breaking down. The I_O rating lists the maximum amount of current the diode can pass in the forward direction. The V_F rating lists the amount of voltage drop in the forward bias direction. This value will be less than 1 volt for most diodes. The V_F rating is of serious concern in power supplies designed to produce high currents.

Some manufacturers produce diodes with their own in-house numbers instead of using the registered 1N system. Motorola often uses the letters MR, MRA, MSD, or MZ to precede the numbers on a diode. Sylvania often used ECG and RCA often uses SK. This different number system doesn't mean the device is not of good quality; it means that the proper information is necessary to find the device's characteristics.

Some diodes are rated to handle only a few milliamperes and others can handle several hundred amperes, figure 4-7. The large stud-mounted type must be properly heat sinked if they are to handle their rated current.

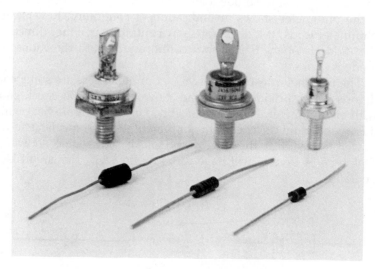

Figure 4-7 Diodes shown in different case styles

Figure 4-8 Schematic symbol of diode

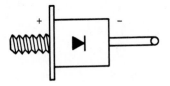

Figure 4-9 Schematic symbol
shows polarity

The connections of the diode are known as the *anode* and *cathode*, figure 4-8. These terms have been carried over from the days of vacuum tubes, when the anode or plate of the vacuum tube was connected to a positive voltage and the cathode or emitter was connected to the negative. If the diode is to be forward biased, the anode must be connected to the positive voltage and the cathode to the negative voltage.

Diodes use some means to identify their polarity. Large power diodes often have the schematic symbol printed on them to show which lead is the anode and which is the cathode, figure 4-9. The small, plastic-case diodes have a line printed on one end. This line represents the line drawn in front of the arrow on the schematic symbol, figure 4-10.

Testing the Diode

Diodes can be tested with an ohmmeter. Since the diode operates like an electric check valve, it should show continuity when the ohmmeter leads are connected to it in one direction but not in the other. To test the diode, connect an ohmmeter across it. Notice if the ohmmeter shows an indication of continuity. Reverse the ohmmeter leads. If it does not show continuity in either direction, the diode is open. If it shows continuity in both directions the diode is shorted.

The ohmmeter test is about 98% accurate. There are some conditions when a diode can be bad, however, and still check good with an ohmmeter. If the diode is breaking down from heat while it is operating in the circuit, the ohmmeter will not furnish enough power to overheat the device and make the trouble show up. This type of trouble will have to be found while the circuit is in operation.

Figure 4-10 Polarity indicated by line

APPLICATIONS

The diode has found many thousands of uses in the home and industry. Two circuits using a single diode will be discussed. The first circuit, figure 4-11, is a dimmer control for a lamp. The switch is a single pole double-throw with a center-off position. When the switch is in the center position, there is no power applied to the lamp. When the movable contact makes connection with the upper stationary contact, full voltage is connected to the lamp and it burns at full brightness. If the movable contact is changed to permit it to make connection with the lower stationary contact, the diode is connected in series with the lamp. Since the diode permits current to flow through it in only one direction, half of the ac waveform is blocked during each cycle. This permits only half the voltage to be applied to the lamp, causing it to burn at half brightness. The advantage of this type of dimmer control is that standard light bulbs can be used. This dimmer circuit does not require the use of the more expensive 3-way bulbs.

The second circuit to be discussed is shown in figure 4-12. This circuit is very similar to the circuit shown in figure 4-11 with the exception of a different type of switch and the addition of a current limiting resistor. This circuit is used to provide a dynamic brake for a small ac induction motor. AC induction motors can be braked by applying direct current to their stator windings. A diode permits current to flow through it in only one direction. Since the current never reverses direction, it is direct current.

The switch used is a single pole double-throw with a center-off position. When the switch is thrown in one direction, it maintains contact. When the switch is thrown in the other direction, it is a momentary contact. The circuit is so constructed that when the

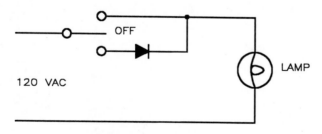

Figure 4-11 Light dimmer control

switch is thrown in the maintained contact position, the ac motor is connected to the power line. When the switch is thrown in the momentary contact position, it connects the diode and the current limiting resistor in series with the ac induction motor. This applies direct current to the stator winding and brakes the motor. The resistor is sized to limit the amount of current flow to a safe value for both the diode and the winding of the ac motor.

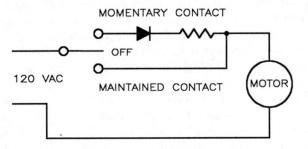

Figure 4-12 Dynamic braking circuit for ac induction motor

UNIT 5
The Light-emitting Diode

The *light-emitting diode* (LED) is a special diode that emits light when current passes through it. The schematic symbol for the LED is shown in figure 5-1. Notice that the arrow points away from the diode. This indicates that light is being given off by the device. The photodiode is turned on by light. The schematic symbol for the photodiode is shown in figure 5-2. The symbols are the same except that the arrow points toward the diode symbol for the photodiode. This shows that the device must receive light if it is to operate.

The light-emitting diode and the junction diode are similar in many respects. Both are rectifiers and will permit current to flow in only one direction. The LED, however, has a higher voltage drop than the junction diode. A silicon junction diode must have about .7 volts to turn it on. The LED must have about 1.7 volts before it can be turned on. This higher voltage drop makes the LED difficult to test with some ohmmeters. An ohmmeter must produce enough voltage to turn the LED on, and many do not have this ability. The best method for testing the LED is to construct a circuit and see if the LED will operate.

When used in a circuit, the LED is generally operated at about 20 milliamperes (ma) or less. For instance, if an LED is to be connected into a 12 volt dc circuit, a current-limiting resistor must be connected in series with the LED, figure 5-3. The resistor needed is figured by the formula: $R = E/I$ $R = 12/.020$ $R = 600$ ohms. The nearest standard 5% resistor value is 620 ohms. Therefore, a 620 ohm resistor would be used as a current-limiting resistor.

When connecting an LED into a circuit, it is necessary to know which lead is the anode and which is the cathode. If the LED is held with the leads facing you, the plastic case will have a flat side beside one of the leads, figure 5-4. The flat side corresponds to the line in front of the arrow on the diode symbol.

Figure 5-1 Light-emitting diode

Figure 5-2 Photodiode

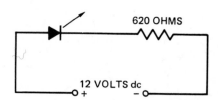

Figure 5-3 Current flow must be limited by a resistor

Figure 5-4 LED polarity

25

Figure 5-5 Seven segment display

The LED has become very popular in electronics since its invention. This is mainly because the LED is inexpensive and has no filament to burn out. LEDs are used as pilot lights on electronic equipment and numerical displays. A very common LED device is the seven segment display which is designed so that seven LEDs can be lighted in different combinations to form any number, figure 5-5.

Seven segment displays are made with one lead of all the LEDs connected together. If this lead is a cathode, then all the cathodes are connected together to form a common terminal. A positive voltage is then connected through a current-limiting resistor to each LED segments which is to be lighted. Seven segment displays are also made with a common anode. With this display, a positive voltage is connected to the anode. The individual segment is connected to negative or ground to light that particular segment.

UNIT 6
Single-phase Rectifiers

A *rectifier* is a device which changes alternating current into direct current. The simplest rectifier is the half-wave rectifier. It gets its name because it can change only one half of the ac voltage applied to it into direct current, figure 6-1.

RECTIFIER OPERATION

The circuit in figure 6-1 contains a single diode connected in series with a current-limiting resistor. Assume a current flow of positive to negative. When the ac voltage applied to the anode of the diode rises in the positive direction, the diode is forward biased and conducts a current flow through the load resistor. If an oscilloscope is connected across the resistor, a voltage shows for that half cycle. When the ac voltage goes in the negative direction, the diode is reverse biased and does not conduct for that half cycle. Since no current flows through the resistor during this half cycle, no voltage is dropped and the oscilloscope indicates 0 voltage. Since only the positive half of the ac wave is conducted, the current through the resistor never reverses its direction, figure 6-2. Therefore, it is direct current. Since this dc voltage turns on and off, it is called pulsating direct current. The half-wave rectifier is generally used only for very low power applications.

The full-wave rectifier is used more often because it can rectify both halves of the ac waveform. There are two types of full-wave rectifiers used for single-phase ac; the *two-diode type* and the *bridge*.

OBJECTIVES

After studying this unit the student should be able to:

- Discuss the operation of single-phase rectifiers
- Connect a two-diode type of single-phase rectifier using discrete electronic components
- Connect a bridge rectifier using discrete electronic components
- Compute the output voltage for different types of rectifiers

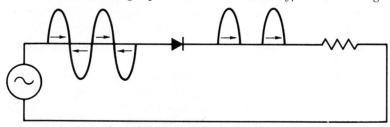

Figure 6-1 Half-wave rectifier

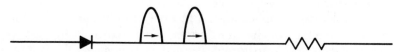

Figure 6-2 Waveform produced by half-wave rectifier

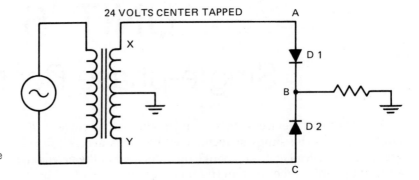

Figure 6-3 Two-diode type of full-wave rectifier

Two-diode Rectifier

The two-diode type of rectifier is more efficient because it has a voltage drop of only one diode. It must be used with a center-tapped transformer, however, figure 6-3.

To understand this rectifier, assume that point X is positive and point Y is negative. If the current flows from positive to negative, it will flow from point X of the transformer to point A of the rectifier. Diode D1 is forward biased, and current flows through D1 to point B. Diode D2 is reverse biased, so the current flows through the load resistor to ground. It returns to the secondary of the transformer through the grounded center tap to point Y which is negative during this half cycle. Current flowed through the load resistor during the half cycle, so a voltage appears across the load resistor.

During the next half cycle of alternating current, point Y is positive and point X is negative. Current flows from point Y of the transformer to point C of the rectifier. Diode D2 is forward biased, so current flows through D2 to point B. Diode D1 is reverse biased now so the current flows through the load resistor to ground. It returns to the secondary of the transformer through the grounded center tap to point X which is negative. Current flowed through the load resistor during this half cycle, so a voltage appears across the resistor.

Current flowed through the resistor during both half cycles of ac and it flowed in only one direction. Therefore, a voltage was developed across the resistor during each half cycle of ac voltage, figure 6-4. Since it flowed in only one direction, the voltage produced was dc. The voltage and current pulsate, but only half as much as the half-wave rectifier.

The ac voltage applied to the rectifier is only half of the full secondary voltage of the transformer. For instance, if the secondary voltage of the transformer is rated at 24 volts ac, the voltage applied to the rectifier is actually 12 volts ac.

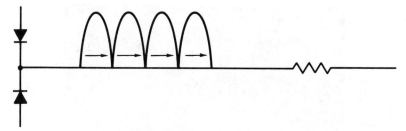

Figure 6-4 Waveform produced by full-wave rectifier

The Bridge Rectifier

The bridge is the other type of full-wave rectifier. It requires four diodes for its construction, but has the advantage of not needing a center-tapped transformer, figure 6-5.

To remember the proper diode connection for a bridge rectifier, think nose to nose and back to back. Notice in figure 6-5 that the two diodes which form the positive terminal of the rectifier have their nose ends connected, and the two diodes which form the negative terminal have their back ends connected.

Assume that point X of the ac source is positive and point Y is negative, figure 6-5. Current flows from X to point A. At point A, diode D4 is reverse biased and D1 is forward biased. The current flows through D1 to point B. At point B, diode D2 is reverse biased, so the current flows through the load resistor to ground. It returns through ground to point D. At point D, both diodes D3 and D4 are forward biased, but since current flows from positive to negative it will flow through diode D3 to point C of the bridge. It then moves to point Y of the ac source which is negative. Since current flowed through the load resistor during this half cycle, a voltage is developed across the resistor.

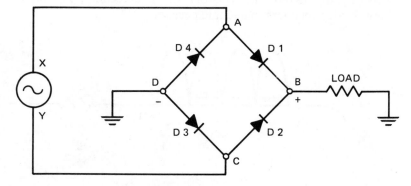

Figure 6-5 Bridge rectifier

Figure 6-6 Bridge rectifiers in a single package
PHOTO COURTESY OF INTERNATIONAL RECTIFIER

Now assume that point Y of the ac source is positive and X is negative. Current flows from point Y to point C of the rectifier. At point C, diode D3 is reverse biased and diode D2 is forward biased. The current flows through D2 to point B. At point B, diode D1 is reverse biased, so the current flows through the load resistor to ground. It flows from ground to point D. At point D, both diodes D3 and D4 are forward biased. Current flows from positive to negative, so the current flows through diode D4 to point A. It then moves to point X which is negative. Since current flowed through the load resistor during this half cycle, a voltage is developed across the resistor.

Notice that the current flow was in the same direction through the resistor during both half cycles, figure 6-6. The waveform produced by the bridge rectifier is the same as the waveform produced by the two-diode type of full-wave rectifier. Figure 6-7 shows bridge rectifiers in a single package.

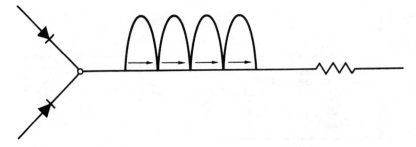

Figure 6-7 Waveform produced by bridge rectifier

Computing Average Value

When ac voltage is rectified to dc, the output voltage of the rectifier is known as the average value. This can be determined by multiplying the peak value (rms × 1.414) of the waveform by .637 or by multiplying the rms value of applied alternating current by .9.

- If an rms voltage of 120 volts is rectified to direct current, the output voltage of the rectifier would be:

  120 volts ac (VAC) × .9 = 108 volts dc (VDC)

- If 120 volts alternating current rms is changed to its peak value it would be 169.68.

  120 × 1.414 = 169.68 and 169.68 × .637 = 108 volts ac

- These values are for a full-wave rectifier. If a half-wave rectifier is used, the average value must be divided by 2.

UNIT 7

The Polyphase Rectifier

In the previous unit, single-phase ac voltage was changed into dc voltage. Industry, however, is operated by three-phase power, so it is necessary to change three-phase alternating current into direct current. There are two types of three-phase rectifiers, the half-wave and the bridge.

THE HALF-WAVE RECTIFIER

The half-wave, three-phase rectifier must be connected to a wye (star) system which has a grounded center tap or a fourth conductor connected to the center tap, figure 7-1.

Notice in figure 7-1 that a diode is connected in series with each phase of the system. The diodes are forward biased when the voltage of each line becomes positive and reverse biased when the voltage becomes negative. As the voltage of each of the three-phase lines goes positive, current flows through the load resistor to ground. It then flows from ground to the center tap of the transformer to complete the circuit. This rectifier has a higher average voltage output and less ripple than a single-phase, full-wave rectifier.

Although the figure shows a half-wave rectifier, it is changing three separate phases which are 120 degrees out of phase with each other into direct current. The rectified dc voltage never falls back to zero volt before another of the three-phase lines begins conducting. Therefore, the dc voltage never reaches the zero reference line as it does with a single-phase, full-wave rectifier. Figure 7-2 (A) and figure 7-2 (B) show the difference between these two rectifiers.

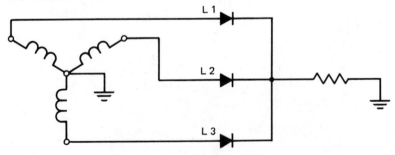

Figure 7-1 Three-phase half-wave rectifier

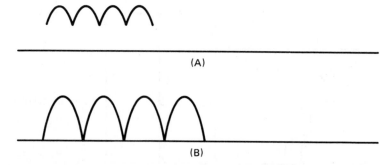

Figure 7-2 (A) Ripple produced by a three-phase half-wave rectifier
(B) Ripple produced by a single-phase full-wave rectifier

THE BRIDGE RECTIFIER

The three-phase bridge rectifier is often used for industrial applications because it has less ripple than the half-wave rectifier and does not require a center-tapped, wye-connected transformer for operation, figure 7-3. It needs only to be connected to three-phase power for operation. Therefore, power can be supplied by either a wye or delta system.

The three-phase bridge type rectifier also has a higher average dc voltage than the three-phase, half-wave rectifier, because the bridge rectifier changes both the positive and negative halves of the ac voltage into dc, figure 7-4.

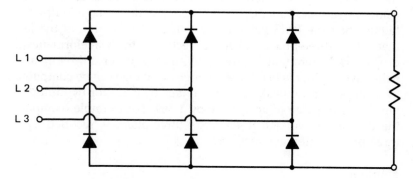

Figure 7-3 Three-phase bridge rectifier

Figure 7-4 Ripple produced by a three-phase bridge rectifier

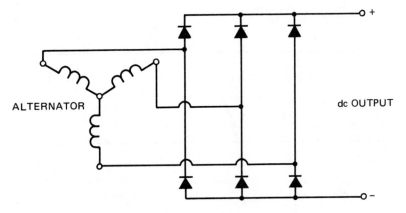

ALTERNATOR

dc OUTPUT

Figure 7-5 Automobile alternator circuit

A very common application of this rectifier is in the alternator of an automobile. Most automobiles use a wye-connected three-phase alternator to supply the power of the charging system. The three-phase ac power produced by the alternator is converted into dc by a three-phase bridge rectifier, figure 7-5.

AVERAGE VOLTAGE CALCULATIONS

The average output voltage for a three-phase rectifier is higher than that for a single-phase rectifier. The reason for this is that there is less ripple when a three-phase rectifier is used. When using three-phase rectifiers, the average dc voltage output is higher than the RMS value of ac voltage. This is due to the fact that when three-phase voltage is rectified, the waveform never drops back to zero volt, figures 7-2A and 7-4. When a three-phase, half-wave rectifier is used, the average dc voltage can be computed by multiplying the peak value of voltage by 0.827, or by multiplying the RMS value of ac voltage by 1.169. For example, compute the dc average voltage if 480 volts three-phase is rectified by a three-phase, half-wave rectifier.

Solution:

In the first method, the peak value of voltage is found by multiplying the RMS value by 1.414.

$$480 \times 1.414 = 678.72 \text{ volts}$$

The peak voltage is now multiplied by 0.827 to find the average dc voltage.

$$678.72 \times 0.827 = 561.3 \text{ volts}$$

The average dc voltage can also be computed by multiplying the RMS value by 1.169.

$$480 \times 1.169 = 561.12 \text{ volts}$$

The slight difference in answers is caused by rounding off values.

The average value of dc voltage can be computed for a full-wave, three-phase rectifier by multiplying the peak value of voltage by 0.955, or by multiplying the RMS value of ac voltage by 1.35. In the following example, the average dc output voltage for a three-phase, full-wave rectifier connected to 480 volts ac is computed.

In the first method the peak voltage will be multiplied by 0.955.

$$678.72 \times 0.955 = 648.12 \text{ volts}$$

The average voltage will now be computed using the RMS value of ac voltage.

$$480 \times 1.35 = 648 \text{ volts}$$

APPLICATIONS

Commercial and industrial applications sometimes require a direct current with very low ripple. Some electronic equipment, such as radio and television transmission equipment, must have a

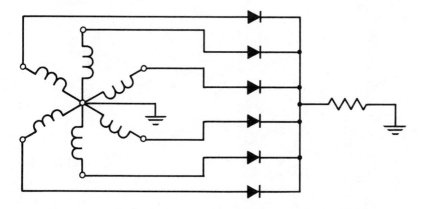

Figure 7-6 Six-phase half-wave rectifier

very smooth direct current to operate. The ripple produced by changing ac voltage into dc voltage can be made much smoother by the use of filters. These will be discussed further in unit 8. The amount of filtering required is partly determined by the amount of ripple the rectifier produces. The three-phase bridge rectifier produces much less ripple than the single-phase bridge rectifier. The three-phase bridge rectifier requires much less filtering, therefore, than a single-phase bridge rectifier of the same power rating.

The amount of ripple produced by a rectifier can become a matter of great importance. For this reason, there are rectifiers that operate on six, twelve, or eighteen phases. These multiple phases are obtained by various transformer connections to a three-phase system. When a multi-phase rectifier is used, it is generally connected as a half-wave rectifier. A six-phase half-wave rectifier has the same ripple as a three-phase bridge rectifier, figure 7-6.

UNIT 8
Filters

In previous units, ac voltage was changed into dc voltage with a rectifier. Regardless of the type of rectifier used, the dc voltage pulsates. The amount of pulsation (ripple) is determined by the type of rectifier used. A half-wave, single-phase rectifier produces the greatest amount of ripple. The three-phase bridge rectifier produces the least. Some types of loads such as dc motors, dc relays, and battery chargers will operate without problems on pulsating (unfiltered) dc, but other electronic loads will not.

PUTTING THE FILTER TO USE

Assume that a radio or tape player that is designed for use in an automobile is operated inside a house. The power in a home is 120 V ac, and the power in an automobile is 12 V dc. The 120 V ac will have to be stepped down to 12 V ac with a transformer, and then rectified into dc, figure 8-1.

Although 12 VDC is now available to operate the radio, it will probably produce an annoying hum. The hum is caused by the pulsations of the dc voltage being applied to the radio. If the radio is to operate properly, the pulsations must be removed.

The pulsations or ripple of the dc voltage can be removed with the proper filter. There are two components used for filtering: a *capacitor* is used to filter the voltage, and an *inductor* or *choke* to filter the current.

OBJECTIVES

After studying this unit the student should be able to:
- Discuss the operation of filters in an electronic circuit
- Discuss the differences between capacitive and inductive filters
- Connect a capacitive filter into a circuit
- Connect an inductive filter into a circuit

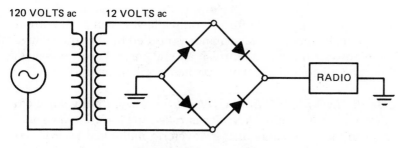

120 VOLTS ac   12 VOLTS ac

RADIO

Figure 8-1 DC power supply

37

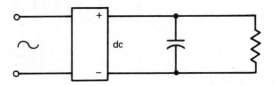

Figure 8-2 Capacitor filter

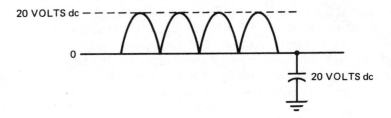

Figure 8-3 Unfiltered dc with a peak value of 20 volts

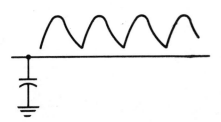

Figure 8-4 Small amount of filtering

The Capacitor as a Filter

When a capacitor is used as a filter, it is connected in parallel with the output of the power supply, figure 8-2. Assume that the rectifier or power supply produces an output voltage which has a peak value of 20 volts, figure 8-3.

As the voltage rises to its peak value of 20 volts, the capacitor charges to 20 volts also. When the dc waveform tries to drop back toward zero, the line voltage becomes less than the 20 volts the capacitor has been charged to. Therefore, the capacitor discharges back into the line in an effort to keep the voltage from decreasing in value, figure 8-4. The amount of filtering that is accomplished is determined by two things:

- the amount of current the power supply must furnish to the load and
- the amount of capacitance connected to the circuit.

If the power supply is to furnish only a few milliamps of current, only a small amount of capacitance is needed to filter the voltage. If the power supply must furnish several amps, however, several thousand microfarads (µf) of capacitance will be needed.

COMPUTING FILTER EFFECTIVENESS: The output voltage of a full-wave (or any other) rectifier is called the *average value* and is equal to the peak value of the waveform multiplied by .637. For instance:

1. Assume an ac voltage of 24 volts (rms) is changed into dc.
2. The peak value of the waveform is 24 × 1.414 = 33.9 VAC.
3. The output voltage of the rectifier is 33.9 × .637 = 21.6 VDC. (The average value of voltage is 21.6 volts.)
4. When the output voltage of a rectifier is filtered, the voltage increases in value. The amount of increase is in proportion to the amount of filtering.
5. The waveform shown in figure 8-5 (A) has been filtered only a small amount. The output voltage of this power supply is only slightly higher than an unfiltered power supply.
6. If more capacitance is added to the circuit, the waveform will begin to appear more like the waveform shown in figure 8-5 (B). The voltage is much higher above the zero reference line than the waveform of figure 8-5 (A). The output voltage of the power supply is higher since the voltage doesn't drop as close to 0 volt.
9. If enough capacitance is added to the circuit to completely filter the ripple, the output waveform appears as a straight line, figure 8-6.

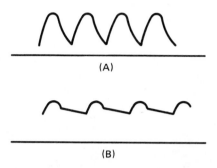

Figure 8-5 (A) Small amount of capacitance (B) Larger amount of capacitance

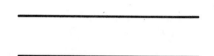

Figure 8-6 DC voltage completely filtered

When all the ripple has been filtered out of the dc voltage, the output voltage of the power supply will be the same value as the peak voltage of the ac applied voltage being rectified. Assume that 24 volts ac rms is connected to a full-wave rectifier. The peak value of the ac waveform is 33.9 volts. If the dc rectified voltage is filtered so that it has no ripple, the dc output voltage will be 33.9 volts.

The Choke Coil as a Filter
The choke coil does for current what the capacitor does for voltage. It is connected in series with the output of the rectifier, figure 8-7.

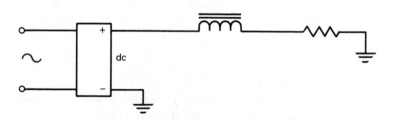

Figure 8-7 Connection of filter choke

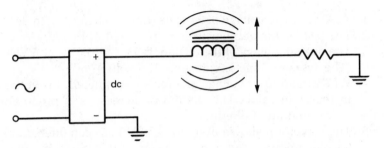

Figure 8-8 Magnetic field increasing with an increase of current

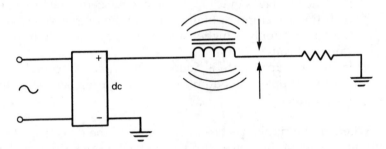

Figure 8-9 Magnetic field inducing current back into the coil

When the current begins to rise from zero toward its peak value, a magnetic field is created around the choke coil, figure 8-8. As the current reaches its peak value, the magnetic field does also. They decrease from their peak value back toward zero in the same way, figure 8-9. As the magnetic field collapses around the choke coil it induces a current back into the coil. This induced current flows in the same direction as the circuit current. Therefore it aids the circuit current by trying to keep a continuous current flow throughout the circuit. The size of the choke coil needed to filter the current is determined by the amount of power the circuit must produce.

COMPARISON OF THE FILTERS

There are various ways to connect capacitors and chokes into a circuit for filtering. A power supply designed for low power application may use a single capacitor filter. A common filtering

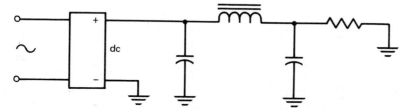

Figure 8-10 "PI" transfer

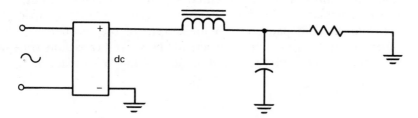

Figure 8-11 Choke input filter

arrangement is the "PI" filter. It gets its name because it resembles the Greek letter "PI," figure 8-10.

Power supplies designed to deliver a larger current generally use a choke input filter, figure 8-11. Choke input means that the choke is connected ahead of the capacitor in reference to the rectifier. This is generally done to keep the surge of current down when the power is turned on. Recall that:

- A discharged capacitor looks like a short circuit to ground when power is first applied to it.
- This can cause a huge current spike which can destroy the diodes in the rectifier.
- By connecting the choke coil ahead of the capacitor, the rate at which the current can rise is limited.
- An inductor opposes a change of current. This opposition to a change of current prevents the high current spike when the power is first turned on.

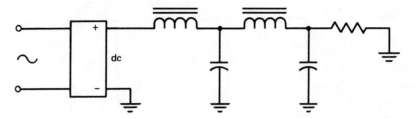

Figure 8-12 Choke input "PI" filter

A common industrial choke input filter is shown in figure 8-12. Notice that this is a "PI" filter with the addition of a choke input.

The schematic drawings of filter networks may show one capacitor connected in parallel with the output of the rectifier. This capacitor on the schematic may actually be several capacitors connected in parallel with each other to increase the total capacitance.

UNIT 9
The Zener Diode

The *zener diode* is a special device designed to be operated with reverse polarity applied to it. When a diode breaks down in the reverse direction, it enters into what is known as the zener region. Most diodes which break down into this zener region are destroyed. The zener diode, however, is designed to be operated in this region without harm to itself.

Notice in figure 9-1 that when the reverse breakdown voltage of a zener diode is reached, the voltage drop of the device remains almost constant regardless of the amount of current flowing in the reverse direction. Since the voltage drop is constant, any device connected in parallel with the zener will have a constant voltage drop even if the current through the load is changing.

Resistor R1 in figure 9-2 is used to limit the total current of the circuit. Resistor R2 is used to limit the current in the load circuit. The value of R1 is less than the value of R2. This is to ensure that the supply can furnish enough current to operate the load. Notice that the supply voltage is greater than the zener voltage. This is necessary or the circuit cannot operate.

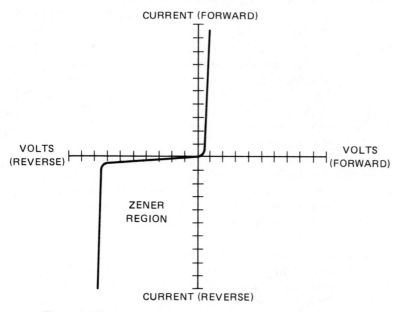

Figure 9-1 The voltage drop is almost constant in the zener region

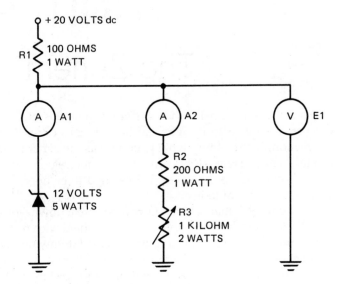

Figure 9-2 The zener diode maintains a constant voltage drop.

Resistor R1 and the zener diode form a series circuit to ground. Since the zener diode has a voltage drop of 12 volts, resistor R1 will have a voltage drop of 8 volts (20 volts − 12 volts = 8 volts). Resistor R1 therefore permits a maximum current flow in the circuit of .08 or 80 ma (milliamperes) 8/100 = .08.

The load circuit which is a combination of R2 and R3 is connected in parallel with the zener diode and the voltage applied to the load is the same as the voltage dropped by the zener. If the zener diode maintains a constant 12-volt drop, the load must have a constant voltage of 12 volts applied to it.

The maximum current which can flow through the load is .06 or 60 ma (12 (V) volts/200 (Ω) ohms = .06 (A) amp). The value of R1 was chosen to ensure that there would be enough current available to operate the load.

When the load is connected as shown in figure 9-2, and resistor R3 is adjusted for 0 ohm, meter A1 indicates a current of 20 ma and meter A2 indicates a current of 60 ma. The current through the zener diode and the load will always add to the value of the maximum current in the circuit permitted by resistor R1 (20 ma + 60 ma = 80 ma). The value of voltage indicated by meter E1 is the same as the zener voltage. If R3 is increased to a value of 200 ohms, the resistance of the load is now 400 ohms (200 + 200 = 400). Meter A1 indicates a current of 50 ma and meter A2 indicates a current of 30 ma. The voltage indicated by meter E1 remains at the zener voltage.

The zener diode makes a very effective *voltage regulator* for the load circuit. Although the current changes, the zener diode forces the voltage to remain at a constant value, and conducts to ground the current not used by the load circuit.

The schematic symbol for a zener diode is shown in figure 9-3. It can be tested with an ohmmeter in the same manner as a common junction diode, provided that the zener voltage is greater than the battery voltage of the ohmmeter.

Figure 9-3
Schematic symbol
of zener diode

APPLICATIONS

A good example of how a zener diode can be used as a voltage regulator can be found in the charging circuit of many motorcycles, figure 9-4. In this circuit, the alternator is used to produce the direct current needed to operate the electrical system of the motorcycle and charge the battery. When a 12-volt battery is fully charged, it will exhibit a voltage of 14 volts across its terminals. If the voltage supplied to the battery becomes greater than 14 volts, there is a danger of overcharging the battery.

In this circuit, the alternator is connected in parallel with the battery. A power zener diode is connected in parallel with both the battery and the alternator. Since electrical components connected in parallel must have the same amount of voltage applied to them, the zener diode will not permit the voltage supplied to the battery to become greater than 14 volts. Notice that there is no current limiting resistor connected in series with the zener diode. The current in this circuit is limited by the internal impedance of the alternator.

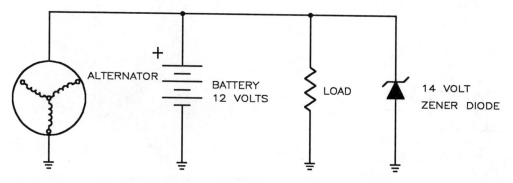

Figure 9-4 Typical voltage regulator circuit for a motorcycle

UNIT 10

The Transistor

OBJECTIVES

After studying this unit the student should be able to:

- Discuss the operation of a transistor
- Name the two types of transistors
- Find the parameters of a transistor in a semiconductor catalog
- Test a transistor with an ohmmeter
- Connect a transistor into an electronic circuit
- Discuss the polarity connections for different types of transistors
- Draw transistor symbols for both NPN and PNP type transistors

The transistor is a semiconductor device made by joining together three layers of p- or n-type material. There are two basic types of transistors: the *NPN* and the *PNP*. The manner in which the layers of semiconductor material are joined together determines the transistor type, figure 10-1.

Germanium and silicon are both used in the production of transistors, but silicon is used more often because of its ability to withstand heat. However, a silicon device requires more voltage to operate. Therefore, when low operating power is required a germanium device may be employed.

IDENTIFYING A TRANSISTOR

Transistors are made with a wide variety of voltage, current, and power ratings. They are made with different sizes and styles of cases. They may also be in a plastic case or a metal case. The type of case generally determines the amount of power the transistor is designed to control, figure 10-2.

Transistors can have:

- voltage ratings from a few to several hundred volts.
- amperage ratings from milliamperes to fifty or more amperes.
- power ratings from milliwatts to several hundred watts.

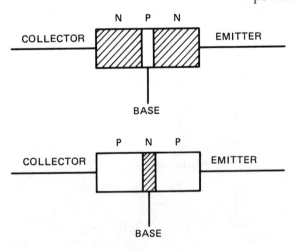

Figure 10-1 Transistors

Figure 10-2 Transistors shown in different case styles

PHOTO COURTESY OF INTERNATIONAL RECTIFIER

NUMBER	MATERIAL	POLARITY	POWER RATING PD	VOLTAGE RATING Vce	CURRENT RATING Ic	FREQUENCY
2N2222	S	N	.5 W	30	150 mw	250 MHz
2N2907	S	P	1.8 W	40	150 mw	200 MHz
2N2928	G	P	150 m	13	2 mw	400 MHz
2N6308	S	N	125 W	350	3 A	5 MHz
2N6280	S	N	250 W	140	20 A	30 MHz

Figure 10-3 Transistor specifications

Additionally, some transistors are designed to operate at a few thousand cycles per second, while others can operate at several hundred million cycles per second.

Transistor ratings can vary greatly and without some means of identifying the transistor, its characteristics may never be known. The most commonly used system is the 2N registry. The transistors listed in this system have a number preceded by 2N, and can be found in a reference book.

Refer to the example in figure 10-3 and see that:

1. The number column gives the 2N number of the transistor.
2. The material column tells if the transistor is made of germanium or silicon.
3. The polarity column indicates if the transistor is an NPN or PNP.
4. The power column tells the amount of heat that the transistor can dissipate. (Note: The 2N6308 and the 2N6280 transistors show power ratings of 125 watts and 250 watts respectively. These power ratings can be obtained only if the transistors have been mounted on proper heat sinks and thermal compound used. The 2N2928 transistor has a power rating of 150 mw which indicates 150 milliwatts or .15 watts.)
5. The voltage rating column indicates the amount of voltage the device can withstand without breaking down.
6. The amperage column indicates the maximum amount of current the transistor can conduct. (Note: A transistor may not be able to hold off maximum voltage while conducting maximum current. The 2N6308 transistor has a power rating of 125 watts, a voltage rating of 350 volts and a current rating of 3 amps. If the transistor were to drop 350 volts while conducting 3 amps of current it would have to dissipate 1050 watts (350 × 3 = 1050).

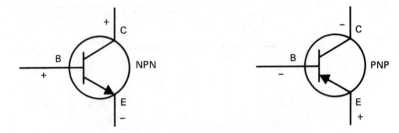

Figure 10-4 Schematic symbols of transistors

7. The frequency column indicates the maximum operating frequency of the transistor. The MHz means megahertz.

Some manufacturers, such as Motorola and RCA, use their own numbering system for some devices. Motorola often uses MJ, MJE, or SDT. RCA uses an SK number to identify some of their transistors. Some equipment manufacturers use their own special numbers to identify components. These numbers mean nothing to anyone except the manufacturer, and in most cases cannot be identified.

Transistor Schematics

Figure 10-4 shows the symbol for the NPN and PNP type of transistor and the polarity markings for each. Notice that the transistors have the base connected to the same polarity as the collector. The arrow on the emitter of the transistor points in the direction of conventional current flow, + to −.

OPERATION OF THE TRANSISTOR

The transistor operates like an electric faucet. The collector is the input and the emitter is the output of the device. The base is

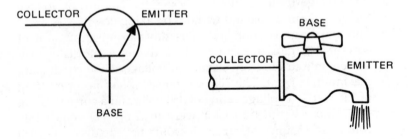

Figure 10-5 Transistors operate like electric faucets.

the control valve, figure 10-5. The current flowing through the base and emitter controls the major current path which is through the collector and emitter of the transistor. A few milliamps of base current can control several hundred milliamps of current through the collector and emitter circuit. The operation of the transistor will be covered in greater detail later in the unit.

Testing the Transistor

Transistors can be tested with an ohmmeter. The test will indicate if the transistor is good or bad. If the polarity of the leads of the ohmmeter is known, it will indicate if the transistor is NPN or PNP. To an ohmmeter, a transistor appears to be two diodes with their anodes (NPN) or their cathodes (PNP) connected.

As shown in figure 10-6, an NPN transistor appears to the ohmmeter as two diodes which have their anodes connected. If the positive lead of the ohmmeter is connected to the base of the transistor, it shows continuity to both the collector and the emitter. If the negative lead of the meter is connected to the base of the transistor, it does not show continuity between the base and the collector or the base and the emitter.

The PNP transistor can be tested by reversing the polarity and connecting the negative lead of the ohmmeter to the base. The meter indicates continuity between the base and the collector and the base and emitter. If the positive lead of the ohmmeter is connected to the base of the PNP transistor it shows no continuity to the collector or the emitter.

The ohmmeter test is considered to be about 98% accurate, but there are some conditions under which this test would not be valid. If the transistor is breaking down under a high voltage the ohmmeter will not be able to supply enough voltage to show the transistor defective. If the transistor is being broken down by heat, the ohmmeter will not supply enough power to show the transistor is bad.

Figure 10-6 Ohmmeter test

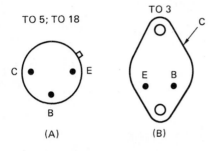

Figure 10-7 Lead identification

Identifying the Leads

The leads of transistors packaged in standard TO3, TO5, or TO18 cases are relatively simple to identify, figure 10-7. Transistors packaged in the TO92, TO220, or TO218 cases can sometimes be difficult because of the position of their leads, figure 10-2. When testing a transistor in the TO5 or TO18 case, hold the transistor upside down. The three leads form a triangle on one side of the case. The little tab on the case is always closest to the emitter lead of the transistor, figure 10-7 (A). The TO3 case transistor leads can be identified by holding the transistor with the leads facing you and down. The lead to the left is the emitter and the lead to the right is the base. The collector is the case of the transistor. Refer to figure 10-7 (B). This can be seen more clearly in figure 10-8, where the matching heat sink holes for the TO3 device are shown.

The plastic case transistors such as the TO92, the TO220, the TO218, etc. can have their leads at any position. The following procedure can be used to identify these transistors.

1. Assume the transistor is an NPN.
2. Assume that one lead of the transistor is the base. Connect the positive lead of the ohmmeter to it.
3. Touch each of the other two leads one at a time. If there is continuity to the other two leads it is the base of the transistor and it is an NPN. If there is no continuity, assume another lead to be the base and repeat the test.
4. If none of the three transistor leads prove to be the base pin, assume the transistor to be a PNP. Repeat the procedure

Figure 10-8 A heat sink for TO3 case devices

PHOTO COURTESY OF AAVID ENGINEERING CO.

using the negative lead of the ohmmeter connected to the base of the transistor.

5. If no lead can be identified as the base of the transistor, the component is probably not a transistor, or it is defective.

6. Once the base lead is identified and it is determined whether the transistor is NPN or PNP, the two remaining leads will have to be identified as to which is the collector and which is the emitter. (Note: The base of the transistor must have the same polarity as the collector in order for the transistor to operate, figure 10-4).

7. Assume that the transistor is an NPN and the base lead has been identified. Assume that one of the two remaining leads is the collector. Connect the positive ohmmeter lead to it. Connect the negative lead to the other pin. The ohmmeter should not indicate continuity.

8. Using a resistor of any value from 10 to 5000 ohms, touch one lead of the resistor to the positive ohmmeter lead and the base of the transistor to the other resistor lead. If the ohmmeter shows conduction, the assumption is correct and the positive ohmmeter lead is connected to the collector of the transistor. If the ohmmeter does not show continuity, reverse the leads and assume the other transistor pin to be the collector.

9. A PNP transistor can be tested in the same manner. Connect the negative ohmmeter lead to the collector and touch the resistor between the negative lead of the ohmmeter and the base pin of the transistor.

Current Flow

A transistor is not the same thing as a variable resistor. A transistor is a current-controlled device. The amount of base current controls the amount of current which flows through the collector and emitter. The transistor follows all the rules of Ohm's Law and its impedance changes, but the change of impedance is caused by a change of current flow through the device. In the circuit shown in figure 10-9 resistor R1 is used to limit the amount of current flow through the collector and emitter of the transistor. Resistor R2 limits the base current if resistor R3 should be adjusted to 0 ohm. Resistor R3 is used to control the amount of current flow to the base of the transistor.

If switch S1 in figure 10-9 is open, no current flows to the base of the transistor, and the transistor will not conduct current through the collector-emitter. In this condition the transistor appears to be an open switch. The voltmeter will show a voltage

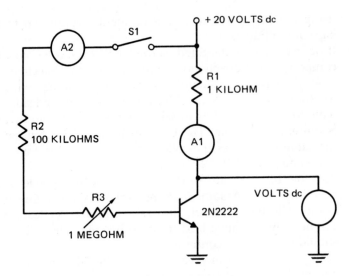

Figure 10-9 Basic transistor circuit

drop across the collector-emitter of 20 volts. Since the transistor has a voltage drop of 20 volts with no current flow, the impedance of the transistor appears to be infinity ($20/0 = \infty$).

Assume that switch S1 has been closed, and resistor R3 has been adjusted to permit a current of 20 microamperes (μa) to flow through the base-emitter of the transistor. The 20 μa of base current permits a current of 2 milliamperes (ma) to flow through the collector-emitter of the transistor. The voltmeter indicates a voltage of 18 volts (V) across the collector-emitter of the transistor. Since the transistor has a current of 2 ma flowing through it and a voltage drop of 18 V, it has an impedance of 9000 ohms (Ω) ($18/.002 = 9000$).

If resistor R3 is adjusted to permit a current flow of 40 μa through the base-emitter of the transistor, more current will be permitted to flow through the collector-emitter. Assume a collector-emitter current of 4 ma. The voltage drop across the collector-emitter of the transistor is 16 V. Since the transistor has a current flow of 4 ma through it and a voltage drop of 16 V across it, the impedance is 4000 Ω ($16/.004 = 4000$).

The base current determines the amount of current flow through the collector-emitter of the transistor. The amount of current flow through the collector-emitter determines the impedance of the transistor. The impedance of the transistor changes, but the current determines the *amount* of change.

APPLICATIONS

A good example of how a transistor can be used to control current flow can be seen in the voltage regulator circuit for many automobiles. The amount of voltage an alternator produces is determined by three factors. These are:

1. The number of turns of wire in the stator.
2. The strength of the magnetic field of the rotor.
3. The speed of rotation of the rotor.

Since the number of turns of wire in the stator is fixed and cannot be changed, the voltage cannot be controlled by changing the number of turns of wire in the stator. Likewise, the speed of rotation of the rotor is determined by the speed of the engine. This prevents controlling the output voltage by controlling the speed of the rotor. The strength of the magnetic field of the rotor, however, is determined by the amount of excitation current supplied to the rotor. If the amount of excitation current is controlled, the output voltage can be controlled.

The function of the voltage regulator is to control the output voltage by controlling the amount of excitation current supplied to the field or rotor. The circuit shown in figure 10-10 is typical of the solid state voltage regulator used by many automobiles. In this circuit, the dc output of the alternator is connected in parallel with

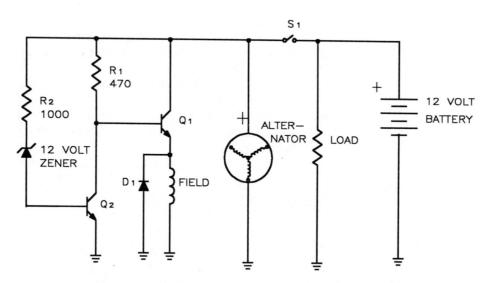

Figure 10-10 Typical voltage regulator circuit for an automobile

the battery and the automobile circuitry (represented here by a LOAD resistor). The voltage regulator is also connected in parallel with the output of the alternator. Transistor Q_1 is connected in series with the field of the alternator. This transistor is used to control the amount of current flow through the field. Diode D_1 is connected in parallel with the field. The function of diode D_1 is to prevent a high-voltage spike from being produced when the power is turned off and the magnetic field collapses around the windings of the field.

Transistor Q_2 is connected to the base of transistor Q_1. Resistor R_1 is used to limit the current flow to the base of transistor Q_1, and through the collector-emitter circuit of transistor Q_2. Transistor Q_2 is used to "steal" the base current from transistor Q_1. A 12-volt zener diode is connected through a current-limiting resistor, R_2, to the base of transistor Q_2.

The circuit operates as follows:

1. When the engine starts and switch S_1 closes, the 12-volt battery supplies current through resistor R_1 to the base of transistor Q_1. This permits transistor Q_1 to start conducting current through its collector-emitter to the field windings of the alternator. As current flows through the field windings, the magnetic field of the rotor increases, causing an increase in the output voltage of the alternator.
2. When the voltage rises above 12 volts, the 12-volt zener diode begins to conduct. This supplies base current to transistor Q_2.
3. Transistor Q_2 begins conducting part of the base current of transistor Q_1 to ground. This causes transistor Q_1 to conduct less current to the field of the alternator, which reduces the output voltage.
4. If the output voltage should drop too low, the zener diode turns off and stops the supply of base current to transistor Q_2. This permits more base current to flow to transistor Q_1, which causes an increase in the output voltage of the alternator.

UNIT 11

The Transistor Switch

The transistor is often used as a simple switch. In this mode the transistor is used as a digital device in that it has only two states, on or off. An advantage of using the transistor in this mode is its ability to handle power.

SWITCH OPERATION

Assume a circuit where there is 100 volts applied and a maximum current flow of 2 amps, figure 11-1. This circuit will consume 200 watts of power when operating (100 V × 2 A = 200 W). If the transistor in this circuit is in the off state, the voltage drop across the collector and emitter is 100 volts at 0 amp. The transistor has to dissipate 0 watt of power (100 × 0 = 0). If the transistor is turned completely on and the base driven into saturation, the voltage drop between the collector and the emitter will be about .3 volt. The transistor must dissipate .6 watt (.3 × 2 = .6). Notice that a transistor used in this manner has the ability to control a large amount of power.

A transistor is generally considered to be turned completely on when the voltage drop between the collector and the emitter is about .7 volt. If the transistor is driven into *saturation*, the voltage drop between the collector and the emitter will be about .3 volt. (Note: a transistor is driven into saturation by furnishing more current to the base of the transistor than is needed for normal operation.) If a transistor requires 10 ma of base current to turn it on, ten ma is considered to be the maximum base current necessary to turn the transistor completely on so that the voltage drop between the collector and the emitter is about .7 volt. If the base of this same transistor was furnished with 15 ma, the transistor would go into saturation. The voltage drop between the collector and the emitter will be about .3 volt.

Voltage Drop

There is reason for concern about a difference in voltage drop of .4 volt (.7 − .3 = .4). Assume a transistor is:

1. connected into a circuit which has a current of 20 amps,
2. has a voltage drop of .7 volt, and

OBJECTIVES

After studying this unit the student should be able to:

- Discuss the use of transistors in a switching application
- Connect a transistor into a circuit and use it in a switching application
- Make measurements of transistor voltage drops using test instruments

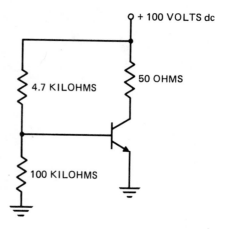

Figure 11-1
Basic transistor circuit

3. will have to dissipate 14 watts of heat (.7 × 20 = 14). If the same transistor:
4. is driven into saturation so that the voltage drop between the collector and the emitter is only .3 volt,
5. it will have to dissipate only 6 watts of heat (.3 × 20 = 6).

There is a significant difference in the amount of heat the transistor must dissipate.

Saturation: When a transistor is driven into saturation, its frequency response is *reduced.* It cannot recover or turn off as quickly as it can under normal conditions. For instance, if a transistor is rated to operate at 1 megahertz (MHz), it may not be able to operate above .5 MHz when driven into saturation. This reduction of a transistor's ability to operate at a high frequency is not generally a problem in industrial circuits, however. Most motor controllers and induction heating equipment seldom operate above about 1000 Hz.

Transistor or Mechanical Switch?

A transistor is used as a switch in place of a regular mechanical switch because it can operate several thousand times a second. It will last longer than a regular mechanical switch even at that switching rate. The versatility of the transistor switch permits it to be operated by a variety of sensing devices such as devices that sense light, sound, temperature, and magnetic induction.

APPLICATIONS

A very common application for this switch is the electronic ignition system of most automobiles. The collector-emitter section of a transistor is connected in series with the primary of the ignition coil. A small induction coil located in the distributor is connected to the base of the transistor. When a magnet is moved past the induction coil, a voltage is induced into the coil. This voltage is used to trigger the base of the transistor which operates the ignition coil, figure 11-2.

Another common use of the transistor switch is the optoisolator found in industry. A light-emitting diode is used to trigger a phototransistor. The LED is controlled by the brains of the circuit, and the transistor is used to operate the power handling part of the circuit, figure 11-3. Optoisolation is used to keep voltage spikes, caused by switching large currents on and off, from "talking" to the electronic control section of the circuit.

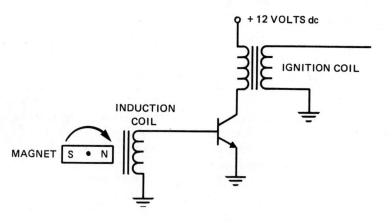

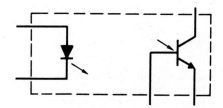

Figure 11-2 Transistor switch operated by magnetic induction

Figure 11-3 Optoisolator

UNIT 12

The Transistor Amplifier

OBJECTIVES

After studying this unit the student should be able to:

- Discuss amplification of electronic signals
- Bias a transistor for use as an amplifier
- Construct a transistor amplifier from discrete components

Transistor amplification is a subject that seems, at first, to be difficult to understand. *Amplification* occurs when a small current through the base and emitter is acted upon by the transistor. This results in a larger current moving through the collector to the emitter. The transistor amplifies a signal produced by the base current.

PRACTICAL APPLICATION

Imagine a hydraulically controlled water valve, figure 12-1. The valve is connected into a water system which has a flow rate of 100 gallons per minute (gpm), and a pressure of 100 pounds per square inch (psi). The valve is controlled by a separate water system which has a flow rate of 1 gpm and a pressure of 1 psi. When 1 gpm of water flows in the control line, the signal causes the main valve to open completely and one hundred gallons of water to flow per minute through the main valve. When the main valve is closed, the pressure across the valve is 100 psi. When the valve is completely open the pressure is 0 psi. A flow of 1 gpm in the control line causes a flow of 100 gpm through the main valve. This is a hydraulic amplifier which has a ratio of 100:1.

The water flowing in the control line is not increased by the main valve. It is the signal from the control line that causes an increase of water flow.

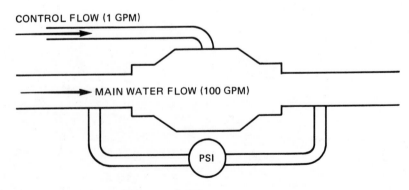

CONTROL FLOW (1 GPM)

MAIN WATER FLOW (100 GPM)

PSI

Figure 12-1 Hydraulic amplifier

Operation as an Amplifier

The transistor operates in a similar manner. The transistor shown in figure 12-2 has a current flow of 100 ma through the collector-emitter and a current of 1 ma through the base. The voltage drop across the transistor will be about .7 volt. If the base current is 0 amp, the collector-emitter will have a current flow of 0 amp and the transistor will have a voltage drop of 20 volts. The base current itself is not increased by the collector-emitter. The 1 ma signal of the base causes 100 ma of collector-emitter current to flow. Therefore, the increase of current flow in the circuit is caused by the 1 ma signal of the base.

When a transistor is to be used as an amplifier, it must be *biased*. Bias means to present or precondition. For example, first consider a transistor amplifier which has not been biased, figure 12-3.

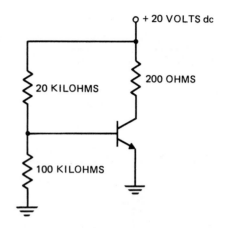

Figure 12-2 Transistor amplifier

Circuit Operation

In this circuit a signal generator supplies an ac signal with a peak voltage of 5 volts. It is assumed that a base-emitter current of 1 ma is required to turn the transistor completely on. Resistor R_1 is used to limit the current flow through the collector-emitter circuit and is the load resistor. Resistor R_2 limits the amount of base-emitter current, and capacitor C_1 blocks any dc component from the signal generator. This ensures that only an ac signal will be applied to the base of the transistor. An oscilloscope is connected across the collector-emitter of the transistor. When the transistor

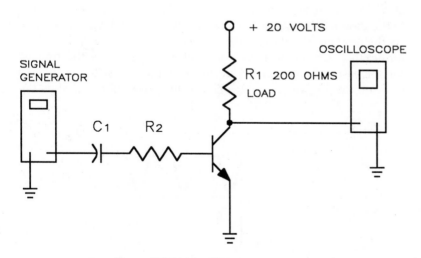

Figure 12-3 Unbiased transistor amplifier

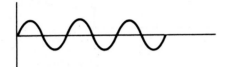

Figure 12-4 AC input signal

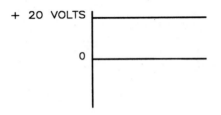

Figure 12-5 20 volts are dropped across the transistor.

is turned completely off, there will be a voltage drop of 20 volts across the collector-emitter. When 1 ma of base-emitter current flows, the transistor will be turned completely on and a voltage drop of about 0.7 volt will appear across the collector-emitter. If the transistor is a linear device, an increase of base-emitter current will produce a proportional increase in collector-emitter current. When the base current is 0, there is no current flow through the collector-emitter and the transistor has a voltage drop of 20 volts.

1. Now assume that a current of 0.25 ma flows through the base-emitter circuit. This produces a current flow of 24 ma through the collector-emitter of the transistor, and a voltage drop of 15.2 volts.
2. If the base current is increased to 0.5 ma, a current of 48 ma flows through the collector-emitter circuit and a voltage drop of 9.7 volts appears across the transistor.
3. If the base-emitter current is increased to 1 ma, the transistor is turned completely on and a current of 98 ma flows through the collector-emitter circuit and the transistor has a voltage drop of 0.7 volts.

The signal generator is producing an ac signal as shown in figure 12-4. When the base current is 0, a voltage of 20 volts will be dropped across the collector-emitter as shown in figure 12-5. As the base-emitter current begins to rise in the positive direction, the impedance of the collector-emitter of the transistor decreases and current begins to flow through the load resistor and transistor. This causes the voltage across the collector-emitter to decrease also. This decrease continues until the sine wave current applied to the base reaches its peak value and begins to fall back toward 0. As the base-emitter current begins to fall back toward 0, the collector-emitter current decreases and the voltage across the transistor increases. The output waveform of the transistor is inverse or opposite the waveform applied to the base. Notice that as the

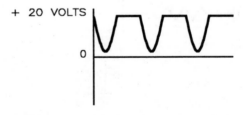

Figure 12-6 Output signal of unbiased amplifier

base current increases, the voltage across the collector-emitter decreases.

When the ac current applied to the base-emitter reaches 0, the transistor is completely turned off and the voltage drop across the collector-emitter is 20 volts. As the base current continues in the negative direction, the transistor cannot turn off more than it is when the base current is 0, so the voltage waveform is not repeated at the output of the transistor. This causes one-half of the waveform to be cut off as shown in figure 12-6.

Biasing the Transistor

In order for the transistor to be able to reproduce the base signal, it must be biased. This is done by furnishing a dc current to the base of the transistor and adjusting the current to a point that the transistor is turned half on, figure 12-7. When resistor R_3 has been adjusted properly, the voltage drop across the collector-emitter will be about 10 volts. This will permit the transistor to be able to turn on as well as turn off. If the base current increases in the positive direction, the transistor will begin to turn on and voltage drop across the collector-emitter will become less than 10 volts. This will continue to happen until the base-emitter current reaches its peak value and begins to turn off. When the base current reaches 0, the voltage across the collector-emitter is back to 10 volts. As the base-emitter current becomes negative, the transistor turns off more and the voltage across the collector-emitter becomes greater than 10 volts.

When the voltage applied to the base of the transistor becomes

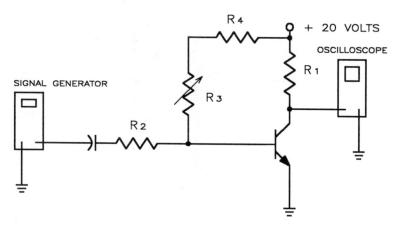

Figure 12-7 Biasing a transistor

Figure 12-8 AC input signal

positive, it aids the positive voltage already being applied to the base by resistors R_3 and R_4. This causes the transistor to turn on more and drop less voltage across the collector-emitter. When the voltage applied to the base becomes negative, it is in opposition to the bias voltage and causes the transistor to turn off and produce a greater voltage drop across the collector-emitter. The voltage waveform applied to the base is shown in figure 12-8, and the voltage waveform produced across the collector-emitter is shown in figure 12-9. The voltage waveform applied to the base has now been reproduced across the collector-emitter although it is inverted.

APPLICATIONS

A good example of how a simple transistor amplifier may be used in an industrial application can be seen in figure 12-10. In this example, a relay is to be controlled by a light beam. A photo diode, D_2, is used to sense the presence of the light beam. A photo diode is a device that will not conduct when it is in darkness, but will conduct when in the presence of light. The photo diode, however, cannot control the amount of current needed to operate the coil of the relay. For this reason, the collector of transistor Q_1 has been connected to the relay coil, and the emitter has been connected to ground. Resistor R_2 is used to ensure the transistor will remain turned off when the photo diode is in darkness. Resistor R_1 limits the amount of current flow through diode D_2 to the base of the transistor when the photo diode turns on. Diode D_1 is used to prevent a spike voltage from being produced when transistor

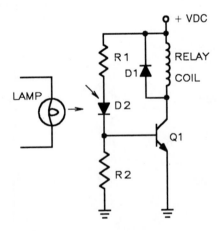

Figure 12-10 Presence of light permits relay to turn on.

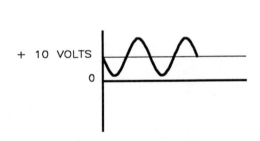

+ 10 VOLTS

0

Figure 12-9 Output signal after biasing

Q_1 turns off and the magnetic field around the relay coil collapses. In this circuit, only a few milliamperes of base-emitter current will flow when the lamp turns on and causes the photo diode to conduct. These few milliamperes are used to control several hundred milliamperes of current through the collector-emitter needed to operate the coil of the relay.

Figure 12-11 illustrates a variation of the circuit shown in figure 12-10. In this circuit, the relay will turn on when the lamp is turned off. When the lamp is turned on, the relay will turn off. When the lamp is turned off, the photo diode has a very high impedance. This permits almost all of the current through resistor R_1 to flow through the base-emitter of the transistor. This permits the transistor to turn on and allow current to flow through the coil of the relay. When the lamp turns on, the photo diode exhibits a very low impedance. This causes almost all the current flowing through resistor R_1 to flow to ground instead of through the base-emitter of the transistor. The transistor turns off and stops the flow of current through the relay coil.

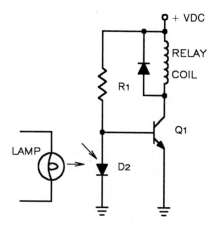

Figure 12-11 The presence of light causes the relay to turn off.

UNIT 13

The Darlington Amplifier

OBJECTIVES

After studying this unit the student should be able to:

- Discuss the operation of a Darlington amplifier
- Compute the gain of a Darlington amplifier circuit
- Construct a Darlington amplifier using discrete components

The Darlington amplifier circuit is one of the most used in industry. It is easily made by connecting the emitter of one transistor to the base of another transistor, figure 13-1.

The Darlington amplifier circuit can have a gain of several thousand. Gain is a comparison of the base current to the collector-emitter current. For instance, if .1 amp of base current is needed to control 1 amp of collector-emitter current, the transistor has a gain of 10 (1/.1 = 10). If .01 amp of base current is needed to produce 1 amp of collector-emitter current, the transistor has a gain of 100 (1/.01 = 100). Gain can vary from one type of transistor to another. It is measured in a term called *beta*. Transistors have a listed beta which is generally given in a transistor reference book. The beta of a transistor is found by dividing the collector-emitter current by the base current. Beta can apply to a circuit as well. If one amp of circuit current is being controlled by 1 ma, the circuit gain is 1000 (1/.001 = 1000).

The Darlington amplifier is used when a high gain is required. Assume that:

- a circuit has a current flow of 10 amps and is to be switched on and off by a power transistor.

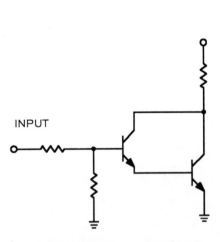

Figure 13-1 Darlington amplifier

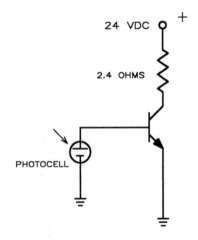

Figure 13-2 A photocell supplies transistor base current.

- the power transistor requires a base current of 10 ma to saturate the transistor.
- the power transistor is to be controlled by a photocell which has a current output of 100 microamperes (μa), figure 13-2.

The sensor device, which is the photocell, cannot supply nearly enough base current to control the power transistor. The photocell can supply only 1/100 the amount of base current needed to drive the transistor into saturation (10 ma/.1 ma = 100). However, if a Darlington amplifier circuit is used, control is easily accomplished, figure 13-3. The photocell is now used to operate a small driver transistor which in turn supplies the base current to the power transistor.

Another way in which the Darlington amplifier is used in industry can be seen in the example circuit in figure 13-4. In this circuit, transistors are used to control a large amount of current to a load. In this circuit, three transistors, Q_1, Q_2, and Q_3 are connected in parallel. These transistors have been connected in parallel to permit them to share the load current necessary to operate the load. Assume that each transistor has a maximum voltage rating of 200 volts and a maximum current rating of 20 amps. This circuit would be able to control a voltage of 200 volts and 60 amps to the load.

When transistors are used in this manner, they must be driven into saturation to produce as small a voltage drop across the collector-emitter as possible. Recall that a transistor is driven into

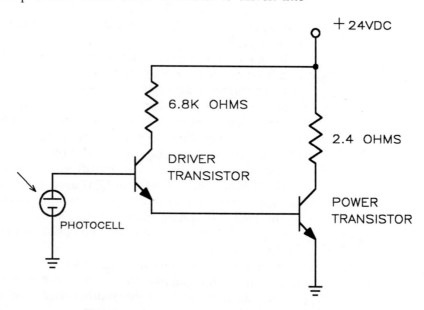

Figure 13-3 Darlington amplifier controlled by a photocell

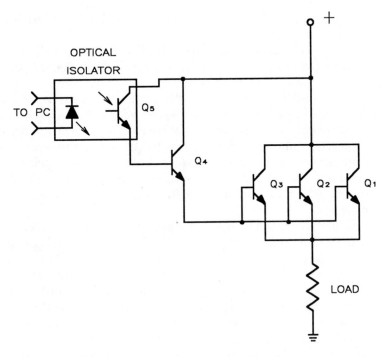

Figure 13-4 Double Darlington amplifier circuit

saturation by supplying it with more base-emitter current than is normally necessary to turn it completely on.

In this circuit, a programmable controller is used to turn the circuit on or off. The programmable controller uses optical isolation to prevent voltage spikes and electrical noise from interfering with the central processor unit. Since the output current of the programmable controller is much less than that required to drive the three power transistors into saturation, another transistor is used to provide base current to the three transistors controlling current flow to the load. This circuit actually contains a double Darlington amplifier. Transistor Q_4 is used as the driver transistor to supply base current to the three transistors controlling current to the load. The photo transistor, Q_5, is the output of the programmable controller. Transistor Q_5 is used as a Darlington drive for transistor Q_4.

A circuit of this type can have a gain of more than a hundred thousand. In this example, the programmable controller furnishes a few milliamps to the light-emitting diode, which in turn can control a current of as much as 60 amps.

UNIT 14
Field Effect Transistors

Field effect transistors (FETs) are so named because they control the flow of current through them with an electric field. There are two basic types of FETs, the junction field effect transistor (JFET) and the metal oxide semiconductor field effect transistor (MOSFET). MOSFETs are often referred to as insulated gate field effect transistors (IGFET). A MOSFET and an IGFET are the same. The first one to be discussed will be the JFET.

JUNCTION FIELD EFFECT TRANSISTORS

JFETs can be divided into two basic types, the n-channel and the p-channel. The schematic symbols for both are shown in figure 14-1. The pins of both the n-channel and p-channel JFETs are labeled *drain, source,* and *gate.* The difference in the two types of JFETs is the polarity of voltage they are connected to. When connecting an n-channel device the drain connects to the more positive voltage and the source and gate connect to a more negative voltage. When the p-channel device is used the drain connects to a more negative voltage and the source and gate connect to a more positive voltage.

JFET OPERATION

A junction field effect transistor is constructed from a piece of n- or p-type material with an electrical field depletion region inserted in it. Figure 14-2 illustrates the construction of an n-chan-

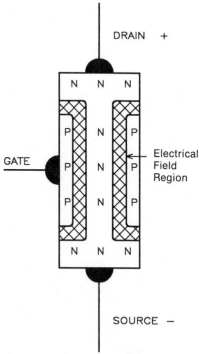

Figure 14-2 Construction of n-channel FET

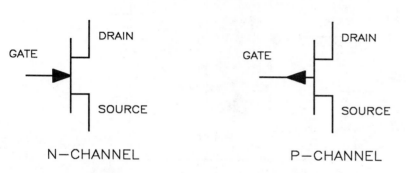

Figure 14-1 Junction field effect transistor symbols

67

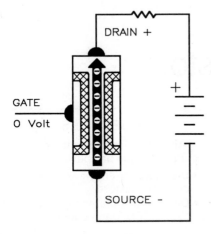

Figure 14-3 Electrons are free to move through the channel.

nel JFET. The channel itself is constructed from a piece of n-type material. Two sections of p-type material form the field depletion region. Both of these two sections are connected to the gate pin, although the connection is not shown in the diagram. If a source of voltage is connected to the source and drain, and no source of voltage is connected to the gate, electrons are free to flow through the channel as shown in figure 14-3. The gate of the JFET controls the current flow through the channel. If the gate of an n-channel JFET is connected to a more positive source of voltage than that connected to the source pin as shown in figure 14-4, the electrical field depletion region becomes smaller and more current is permitted to flow. This, however, is not the manner in which the gate of a JFET is connected. If the JFET were to be connected in this manner, a forward p-n junction would be formed and the device could be damaged. Figure 14-5 illustrates what happens when the gate is connected to a voltage source more negative than the voltage connected to the source pin. If the negative voltage connected to the gate becomes high enough, the flow of current through the channel will be stopped completely by the expanding electrical field region. The electrical field region is used to "pinch off" the electron flow through the channel. Notice that the amount of current flow through the channel is dependent on the amount of *voltage* applied to the gate, not an amount of *current* flow through a base. For this reason, the JFET is said to be a voltage-operated device. In fact, junction field effect transistors operate almost identically to old-style vacuum tubes.

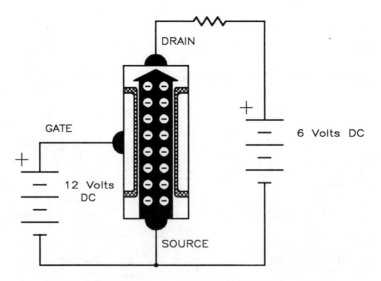

Figure 14-4 A positive gate voltage permits more current to flow.

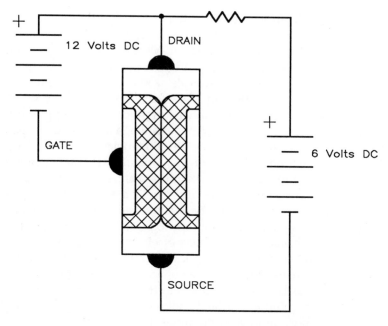

Figure 14-5 Enough negative gate voltage can stop the flow of current.

JFET IMPEDANCE

One of the great advantages of any field effect transistor is the impedance of the gate circuit. The JFET is normally operated with the voltage connected to the source and gate junction reverse biased. This means that the impedance of the gate circuit is that of a reverse biased diode. This impedance can vary from one device to another, but it is generally in the range of about 20,000 MΩ. This means that a voltage of 20 volts between the gate and source will result in a current flow of about one nano-amp (1×10^{-9}). For this reason, field effect transistors are generally considered to require no gate current when operating.

MOSFETs

MOSFETs differ from JFETs in that there is no p-n junction structure. The gate of the MOSFET is insulated from the channel by a layer of silicon dioxide (SiO_2). There are two types of MOS-FETs, the depletion-enhancement MOSFET (DE-MOSFET) and the enhancement-only MOSFET (E-MOSFET). Both of these types can be n-channel or p-channel. Since there is an insulator separating the gate from the channel, the input resistance of a MOSFET is much greater than that of a JFET. This is generally about 100 times greater, which gives the MOSFET an input resistance of about two

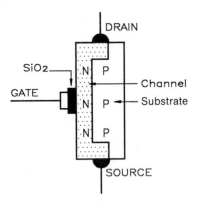

Figure 14-6 N-channel DE-MOSFET

billion ohms. Typical source to gate current is in the region of 10 pico-amps (10×10^{-12}) with a voltage of 20 volts. The basic structure arrangement for an n-channel DE-MOSFET is shown in figure 14-6. The following is an explanation of the operation of an n-channel device. A p-channel device would operate in the same manner with opposite polarity. DE-MOSFETs operate by depletion or enhancement of the channel current carriers as opposed to electrical field depletion. If there is no voltage applied to the gate pin, current is free to flow through the channel as it is in the JFET. When an n-channel DE-MOSFET is operated in the depletion mode, a voltage more negative than that connected to the source is applied to the gate. The negative voltage applied to the gate repels conducting electrons in the channel and leaves positive ions in their place. The electrons are repelled into the p-type substrate region of the device. This *depletion* of electrons reduces the conductivity of the channel, causing less current to flow through the channel. The greater the negative voltage applied to the gate, the greater the depletion of n-channel electrons. If the negative voltage becomes high enough, the n-channel electrons become totally depleted and current flow through the channel stops.

If the DE-MOSFET is operated in the enhancement mode, a voltage more positive than that applied to the source is applied to the gate. This positive voltage attracts electrons from the p-type substrate region into the n-channel region. This increases or *enhances* the conductivity of the n-channel region, which permits more current to flow. Since the gate of the MOSFET is separated from the channel by an insulator and is not a p-n junction, it can be operated with either a positive or negative voltage connected to the gate.

The schematic symbols for both n-channel and p-channel DE-MOSFETs are shown in figure 14-7. Notice the gate is shown not

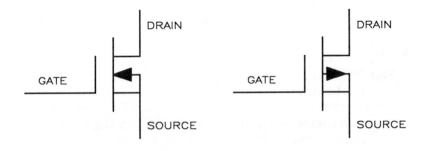

N—CHANNEL DE—MOSFET   P—CHANNEL DE—MOSFET

Figure 14-7 Schematic symbols of DE-MOSFETS

connected to the channel. The symbol also shows the substrate as being connected to the source. This is generally, but not always, true.

E-MOSFETs

Enhancement-only MOSFETs are structured differently from DE-MOSFETs. The structure of an E-MOSFET is shown in figure 14-8. Notice that the E-MOSFET does not contain an actual channel as the DE-MOSFET does. The p-type substrate extends to the silicon dioxide insulation layer. If a voltage is connected to the drain and source and no voltage is connected to the gate, there will be no current flow through the device. In order for current to flow through the device, a positive voltage must be applied to the gate. This causes electrons to be attracted from the p-type substrate and form a conducting channel beside the insulator as shown in figure 14-9. The higher the voltage applied to the gate, the more conductive the device becomes. The schematic symbols for both n-channel and p-channel E-MOSFETs are shown in figure 14-10. The schematic symbol uses a broken line to indicate there is no physical channel.

TESTING FETs

To test a field effect transistor, it must be known if it is a JFET or a MOSFET, and if it is n-channel or p-channel. JFETs can be tested with an ohmmeter. JFETs appear to an ohmmeter to be two

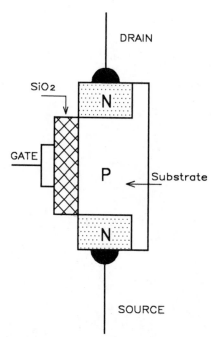

Figure 14-8 Structure of an n-channel E-MOSFET

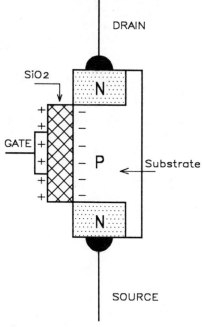

Figure 14-9 Electrons create a channel in the E-MOSFET.

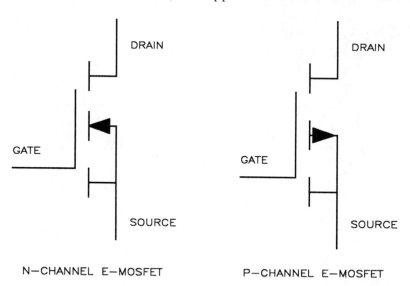

N—CHANNEL E—MOSFET   P—CHANNEL E—MOSFET

Figure 14-10 Schematic symbols of E-MOSFETs

diodes connected together with a resistance connected in parallel with them. The polarity of the diodes is determined by whether the device is n-channel or p-channel. Figure 14-11 shows a schematic diagram of what a JFET should look like to an ohmmeter. When checking an n-channel JFET, if the positive lead of the ohmmeter is connected to the gate, a diode junction should be indicated between gate to source and gate to drain. If the meter leads are reversed, there should be no connection between gate to source or gate to drain. If the ohmmeter leads are connected between the source and drain, some amount of resistance should be indicated. The amount of resistance indicated will depend on the FET. P-channel JFETs can be tested by connecting the negative lead of the ohmmeter to the gate lead and checking for a diode junction between gate to source and gate to drain.

Checking MOSFETs with an ohmmeter is difficult at best. MOSFETs generally come packaged in conducting material that keeps the lead shorted together. This prevents a static charge from building up and damaging the device. MOSFETs can sometimes be damaged just by touching them with a finger. They can be tested with a low voltage ohmmeter set on its highest resistance scale. If the device is a DE-MOSFET, there should be some continuity between the source and drain, but no continuity between the gate to source or gate to drain. If the device is an E-MOSFET, there should be no continuity indicated between any of the pins.

CONNECTING A FET

As stated previously, the gate of a JFET must be connected to a voltage more negative than the source. Figure 14-12 illustrates the use of a separate power supply connected to the gate of the FET

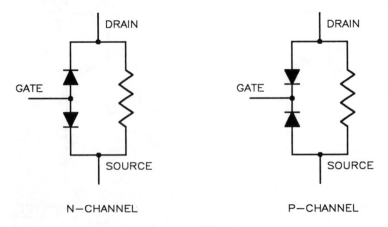

Figure 14-11 Checking JFETs with an ohmmeter

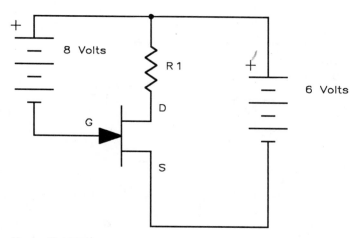

Figure 14-12 The gate must be more negative than the source.

to provide a higher negative voltage than that connected to the source. Most electronic circuits do not employ two separate power supplies, however. To overcome this problem, a resistor is placed in series with the source pin. This forces the source to be at a voltage that is more positive than the gate, figure 14-13. If the source is at a voltage more positive than the gate, the gate must, therefore, be at a voltage more negative than the source. The amount of voltage difference between the source and gate is determined by the amount of current flow through the circuit and the resistance of resistor R_2. If a variable resistor is used in this position, the voltage between source and gate can be adjusted. This permits the current flow through the FET to be controlled by adjustment of the variable resistor.

APPLICATIONS

Field Effect Transistors have found many uses in today's technology. In the circuit shown in figure 14-14, a FET is used as a simple timer. To understand the operation of the circuit, first analyze what happens when switch S_1 is connected to the TIME position. In this position, the switch is open and the gate of the FET is connected to ground through resistors R_1 and R_2. This permits the transistor to turn on and the light-emitting diode glows. When the switch is moved to the RESET position, the gate of the FET is connected to a positive voltage, which turns the transistor off. Capacitor C_1 is charged to the value of the applied voltage. When the switch is moved back to the TIME position, the capacitor

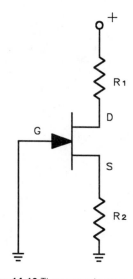

Figure 14-13 The source is more positive than the gate.

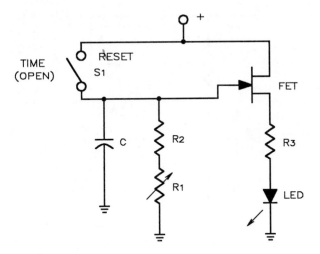

Figure 14-14 FET timer circuit

Figure 14-15 Digital Volt Multimeter

begins to discharge through resistors R_1 and R_2. When the voltage across the capacitor has become low enough, the FET begins to turn on again. Since the transistor turns on at a gradual rate, the LED will begin to glow faintly and brighten as the capacitor discharges and permits the FET to turn on more.

The amount of delay time is determined by the size of capacitor C_1 and the sum of resistors R_1 and R_2. Resistor R_2 is used to limit current in the event resistor R_1 should be set for zero ohm. Resistor R_3 limits current flow through the FET and the light-emitting diode.

One of the most common applications for field effect transistors are digital multimeters, figure 14-15. The circuit shown in figure 14-16 is the basic schematic of a multirange voltmeter. The input impedance of the meter will be 10 meg. ohms regardless of the setting of the voltage range switch. The FETs have taken the place of vacuum tubes, which were used to operate high-impedance meters for many years.

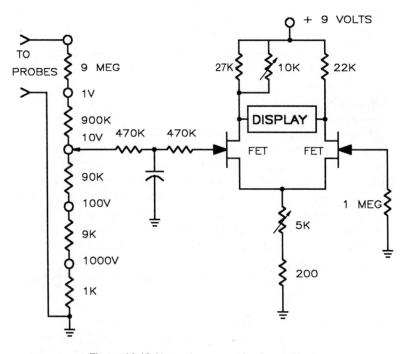

Figure 14-16 Alternating current fundamentals

UNIT 15

Current Generators

OBJECTIVES

After studying this unit the student should be able to:

- Discuss the operation of a current generator
- List common uses for current generator circuits
- Construct a constant current generator using a field effect transistor
- Construct a constant current generator using a junction transistor

A current generator circuit is used to produce an output that is sensitive to an amount of current flow as opposed to some amount of voltage. Current generators are found in various applications. One of the more common uses for the current generator is found in digital ohmmeters. Many digital ohmmeters measure the value of a resistor by passing a known amount of current through it and then measuring the voltage drop across the resistor. In figure 15-1, a current generator is connected to an unknown value of resistance. Now assume that the current generator is producing a constant current of 1 ma (0.001), and that the voltmeter connected across the unknown value of resistance is indicating a voltage of 10 volts. By using Ohm's Law, it can be seen that the resistor has a value of 10,000 ohms.

$$R = \frac{E}{I}$$

$$R = \frac{10}{0.001}$$

$$R = 10,000 \text{ ohms}$$

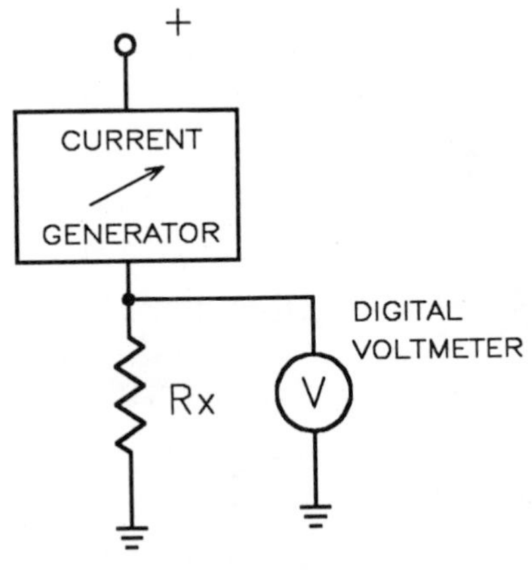

Now assume that the value of resistance has been changed and the voltmeter indicates a voltage of 3.6 volts. The resistor has a value of 3600Ω. It is a simple matter of multiplying any value indicated by the voltmeter by 1000 to indicate the proper amount of resistance.

If the resistance value should become high enough that the circuit cannot furnish the required voltage, the current generator is changed to produce a lower value of current and the multiplication factor is changed. For example, assume the current generator has been changed to produce a constant current of 10 μa (0.000010). If the voltmeter connected across the unknown resistance indicates a value of 2.7 volts, the resistor has a value of 270,000 ohms. The amount of voltage indicated by the voltmeter is now multiplied by a factor of 100,000. Almost any range of resistance can be measured by using the proper amount of current and the correct multiplication factor.

Figure 15-1 The voltage drop across the resistor is proportional to the amount of current flow.

ANALOG SENSORS

Another common use of the current generator is found throughout industry in the use of analog sensing devices. These devices are used to sense temperature, pressure, humidity, etc. They are sensors that are designed to operate between some range of settings, such as 50 to 300 C°, or 0 to 100 P.S.I. These sensors are used to indicate between a range of values instead of just operating in an on or off mode. An analog pressure sensor designed to indicate pressures between 0 and 100 P.S.I. would have to indicate when the pressure was 30 P.S.I., or 50 P.S.I., or 80 P.S.I. It would not just indicate whether the pressure had reached 100 P.S.I. or not. There are several ways that a pressure sensor of this type can be constructed. One of the most common methods is to let the pressure sensor operate a current generator that produces currents between 4 and 20 milliamperes. It is desirable for the sensor to produce a certain amount of current instead of a certain amount of voltage because it eliminates the problem of voltage drop on lines. For example, assume a pressure sensor is designed to sense pressures between 0 and 100 P.S.I. Also assume that the sensor produces a voltage output of 1 volt when the pressure is 0 P.S.I. and a voltage of 5 volts when the pressure is 100 P.S.I. Since this is an analog sensor, when the pressure is 50 P.S.I., the sensor should produce a voltage of 3 volts. This sensor is connected to the analog input of a programmable controller, figure 15-2. The analog input has a sense resistance of 250Ω. If the wires between the sensor and the input of the programmable controller are short enough so that there is almost no wire resistance, the circuit will operate without problem. Since the sense resistor in the input of the programmable controller is the only resistance in the circuit, all of the

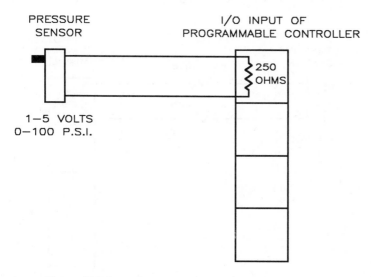

Figure 15-2 A pressure sensor with an output of 1 to 5 volts

output voltage of the pressure sensor will appear across it. If the pressure sensor produces a 3-volt output, 3 volts will appear across the sense resistor.

If the pressure sensor is located some distance away from the programmable controller, however, the resistance of the two wires running between the pressure sensor and the sense resistor can cause inaccurate readings. Assume that the pressure sensor is located far enough from the programmable controller so that the two conductors have a total resistance of 50Ω, figure 15-3. This means that the total resistance of the circuit is now 300Ω (250 + 50 = 300). If the pressure sensor produces an output voltage of 3 volts when the pressure reaches 50 P.S.I., there will be a total current flow in the circuit of 10 ma (0.010 amp) (3/300 = 0.010). Since there is a current flow of 10 ma through the 250Ω sense resistor, a voltage of 2.5 volts will appear across it. This is substantially less than the 3 volts being produced by the pressure sensor.

If the pressure sensor is designed to operate a current generator with an output of 4 to 20 ma, the resistance of the wires will not cause an inaccurate reading at the sense resistor. Since the sense resistor and the resistance of the wire between the pressure sensor and the programmable controller form a series circuit, the current must be the same at any point in the circuit. The pressure sensor will produce an output current of 4 ma when the pressure is 0 P.S.I. and a current of 20 ma when the pressure is 100 P.S.I. When the pressure is 50 P.S.I., it will produce a current of 12 ma. When a current of 12 ma flows through the 250Ω sense resistor, a voltage of 3 volts will be dropped across it. Since the pressure sensor

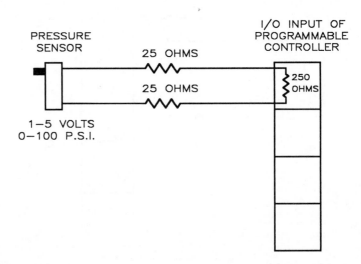

Figure 15-3 Long runs of wire have the effect of adding resistance to the lines.

produces a certain amount of current instead of a certain amount of voltage with a change in pressure, the amount of wire resistance between the pressure sensor and programmable controller is of no concern.

CONSTRUCTING A CURRENT GENERATOR

A current generator can be constructed using either a field effect transistor or a junction transistor. The field effect transistor circuit will be discussed first. The schematic diagram of a current generator using a JFET is shown in figure 15-4. Resistor R_1 is connected between the source and gate. It is also connected in series with the load resistor. The value of R_1 determines the amount of current in the circuit. As current flows through the circuit, a voltage drop is produced across resistor R_1. This voltage drop causes the source of the JFET to become more positive than the gate, which forces the gate to begin reducing current flow through the drain–source. If the current flow through resistor R_1 becomes too low, the voltage drop becomes less and the gate permits more current to flow. The amount of load resistance can now be changed and the current will remain constant within the limits of the circuit.

A pnp junction transistor can also be used to construct a current generator. This circuit is shown in figure 15-5. In this circuit, resistor R_1 is used to determine the amount of current flow through the emitter–collector and load resistor. The zener diode is used to maintain a constant voltage between the base and emitter. In order for current to flow in this circuit, the voltage applied to the emitter must become about 0.7 volts more positive than the voltage applied to the base. Since the zener diode is connected in parallel with resistor R_1, the voltage drop across R_1 must be the same as the zener voltage plus the 0.7 volt needed to operate the transistor. The value of resistor R_1 determines the amount of current that must flow to produce that voltage drop. The value of the load resistor can be changed and the current flow will remain the same as long as the limits of the circuit are not exceeded.

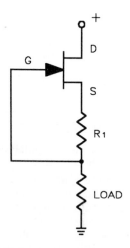

Figure 15-4 A JFET used as a constant current generator

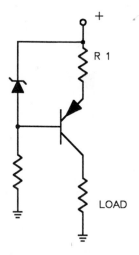

Figure 15-5 A junction transistor used to construct a constant current generator

UNIT 16

The Unijunction Transistor

The unijunction transistor (UJT) is a special device which has two bases and one emitter. It is a digital device because it has only two states, on or off. It is generally classified with a group of devices known as thyristors, which includes silicon-controlled rectifier (SCR), triac, programmable unijunction transistor (PUT), diac, and the UJT.

CONSTRUCTION OF THE UJT

The unijunction transistor is made by combining three layers of semiconductor material as shown in figure 16-1. One of the most common UJTs used in industry is the 2N2646. The schematic symbol with polarity connections and the base diagram is shown in figure 16-2.

Current Flow

The UJT has two paths through it for current flow. One path is from base #2 (B2) to base #1 (B1). The other is through the emitter and B1. In its normal state, there is no current flow through the emitter and B1 path. When the voltage applied to the emitter becomes about 10 volts more positive (higher) than the voltage applied to B1, the UJT turns on. Increased current flows through the B1-B2 path and from the emitter through B1. Current continues to flow through the UJT until the voltage applied to the emitter drops to a point where it is only about 3 volts higher than the voltage applied to B1. At this point, the UJT will turn off until the voltage applied to the emitter again becomes about 10 volts higher

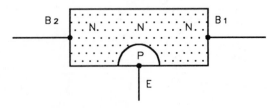

Figure 16-1 Structure of a unijunction transistor

than the voltage applied to B1. The unijunction transistor is generally connected into a circuit similar to the one shown in figure 16-3.

Charge Time

The variable resistor controls the rate of charge time of the capacitor. When the capacitor has been charged to about 10 volts, the UJT turns on and discharges the capacitor through the emitter and B1. When the capacitor has been discharged to about 3 volts, the UJT turns off and permits the capacitor to begin charging again. By varying the resistance connected in series with the capacitor, the amount of time needed for charging the capacitor can be changed, thereby controlling the pulse rate of the UJT, ($T = RC$).

Output Pulse

The unijunction transistor can furnish a large output pulse, because the pulse is produced by the discharging capacitor, figure 16-4. This large pulse is generally used for triggering the gate of an SCR.

The pulse rate is determined by the amount of resistance and capacitance connected to the emitter of the UJT. However, the amount of capacitance which can be connected is limited by the size of the UJT. For instance, the 2N2646 UJT should not have a capacitor larger than 10 microfarads (μf) connected to it. If the capacitor is too large, the UJT cannot handle the current spike produced by the capacitor, and can be damaged.

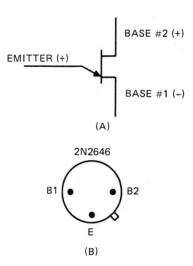

Figure 16-2 (A) Schematic symbol and (B) base diagram

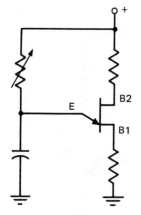

Figure 16-3 Common unijunction transistor circuit

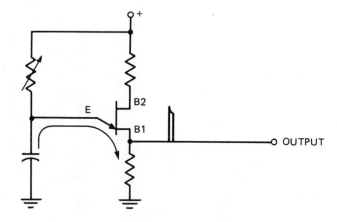

Figure 16-4 Pulse produced by capacitor discharge

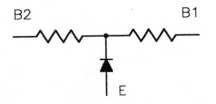

Figure 16-5 Testing a UJT with an ohmmeter

TESTING THE UJT

The unijunction transistor is tested with an ohmmeter in a manner similar to a common junction transistor. The UJT appears to the ohmmeter as a connection of two resistors connected to a diode as shown in figure 16-5. When the positive lead of the ohmmeter is connected to the emitter, a diode junction from the emitter to B2 and another connection from the emitter to B1 is seen. If the negative lead of the ohmmeter is connected to the emitter of the UJT, no connection is seen between the emitter and either base.

APPLICATIONS

A circuit using a unijunction transistor is shown in figure 16-6. This circuit is used as an equipment ON indicator. When some piece of equipment is in operation, a positive voltage is provided at the + input. Capacitor C_1 is charged through resistor R_1. When the voltage of the capacitor exceeds the threshold voltage of the UJT, it turns on and discharges capacitor C_1 to resistors R_2 and R_3. Some of the current flows through resistor R_2 to ground, and some flows through resistor R_3 to the base-emitter of the transistor. This permits the transistor to turn on and provide a current path through the light-emitting diode and resistor R_4. The transistor will remain turned on until capacitor C_1 discharges to a low enough voltage to permit the UJT to turn off. At that point capacitor C_1 begins charging and the process starts over again. As long as a positive voltage is provided at the + input, the light-emitting diode will flash on and off to indicate that the equipment is in operation.

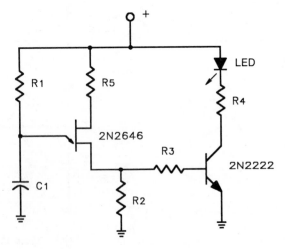

Figure 16-6 Equipment on indicator

UNIT 17

The SCR in a DC Circuit

The silicon-controlled rectifier (SCR) is made by bonding four layers of semiconductor material together to form a p-n–p-n junction, figure 17-1.

The SCR is called a thyristor because its characteristics are similar to the gas-filled thyratron tube used in industrial electronics for many years. Since the invention of the SCR other devices with similar operating characteristics have been invented and are also listed as thyristors. Today there is an entire family of devices known as thyristors.

OBJECTIVES

After studying this unit the student should be able to:

- Discuss the operation of an SCR when connected to a DC circuit
- Draw the schematic symbol for an SCR
- Test an SCR with an ohmmeter
- Connect an SCR in a DC circuit

IDENTIFYING THE SCR

SCRs are known as the workhorses of industrial electronics because of their ability to control hundreds of volts and amps. However, they are also found in smaller sizes with ratings of less than 1 amp at only a few volts. SCRs are also found in many different sizes and styles of cases.

Notice in figure 17-2 that SCRs can be found in many of the same case styles as transistors. Without some means of identifying

ANODE
P
N
P
N
CATHODE
GATE

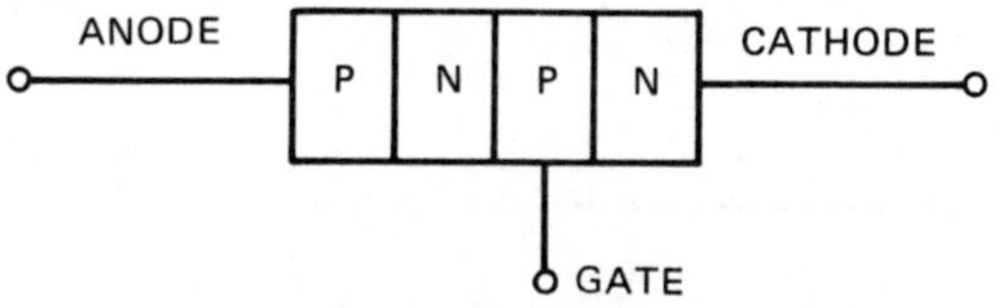

Figure 17-1 Structure of a silicon-controlled rectifier (SCR)

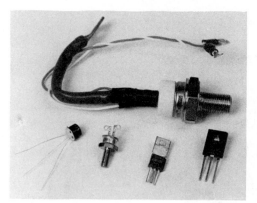

Figure 17-2 Different case sizes and styles for SCRs

the component there is no way of knowing what a device is by looking at the case. SCRs are numbered in a manner similar to transistors. These include:

- a registered 2N number. They can be identified in a semiconductor index.
- a manufacturers' special number. They can be identified with the proper information.
- numbers by equipment manufacturers. They can't be identified by anyone other than the manufacturer.

OPERATING CHARACTERISTICS

To understand the operating characteristics of the SCR, it is necessary to first understand how it operates when connected in a dc circuit. The schematic symbol for an SCR is shown in figure 17-3.

The SCR operates very similar to a relay which is controlled by a set of pushbuttons and is latched with holding contacts, figure 17-4. In this circuit the start button is used to turn relay coil CR on. Once the coil is turned on, the holding contacts, operated by coil CR, close and maintain a current path around the normally open pushbutton, figure 17-5.

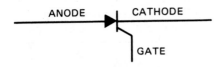

Figure 17-3 Schematic symbol of the silicon-controlled rectifier (SCR)

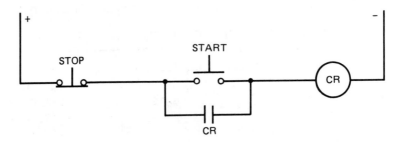

Figure 17-4 Start-stop pushbutton circuit

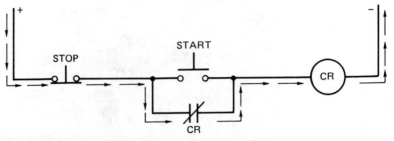

Figure 17-5 Contacts maintain circuit

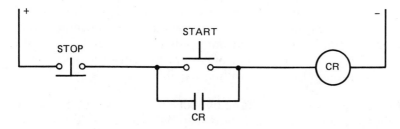

Figure 17-6 Current path broken by stop button

Once the holding contacts have closed, current continues to flow to coil CR even if the start button is released and the push-button is again opened. The start button can control when the relay coil is turned on, but once the relay is energized, the start button has no more control over the circuit.

In order to turn the relay off, the normally closed stop button must be pushed to open the circuit to the coil, figure 17-6. After the relay coil has been deenergized, CR contacts open and the stop button can be released.

Operation of the SCR

The SCR operates in a manner similar to this relay circuit. The SCR in figure 17-7 will not conduct until switch S2 is closed and current flows through the gate-cathode circuit. When the gate energizes, the SCR conducts through the anode-cathode circuit. After the SCR is turned on by the gate, switch S2 can be opened (removing current from the gate), and the SCR continues to conduct. When switch S1 is opened, current flow through the anode-cathode circuit is broken, and the SCR turns off. Switch S1 can now be closed and the SCR will remain turned off until the gate is again used to turn it back on. Notice that the gate operates similar to the start button in the relay circuit. It can turn the SCR on, but not off.

GATE CURRENT: The amount of gate current needed to turn the SCR on varies from one device to another. SCRs designed to control small amounts of power require a gate current of only a few

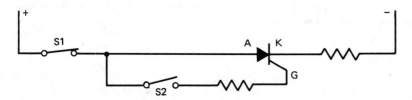

Figure 17-7 Silicon-controlled rectifier operates similar to a start-stop pushbutton circuit

microamps, while SCRs designed to control large amounts of power require a gate current of several hundred milliamps. However, the amount of gate current needed to fire the SCR is only a small fraction of the amount of current the device is designed to handle through the anode-cathode circuit.

HOLDING CURRENT: The amount of current flowing through the anode-cathode circuit needed to keep the SCR turned on is called the *holding current.* Assume a certain SCR has a holding current of 100 milliamps. As long as current flowing through the anode-cathode circuit is above 100 milliamps, the device remains on. If the current drops below 100 milliamps, it turns off. As with gate current, the amount of current needed to keep an SCR turned on varies from one device to another. A general rule is, the more power an SCR is designed to control the higher its holding current will be.

Controlling Power

The SCR is the workhorse of industrial electronics because it can control large amounts of power. This is because it has only two states of operation, completely on or completely off.

In the off state of figure 17-8, the full 200 volts is dropped across the SCR. Since there is no current flowing through the circuit, there is no power being dissipated in the form of heat (200 × 0 = 0). When the gate receives a current pulse through switch S1, the SCR turns on and about 20 amps of current flows through the circuit (200/10 = 20). In the on state, the SCR has a voltage drop of about 1 volt. In this circuit it has to dissipate about 20 watts of heat (20 × 1 = 20), and will have to be heat sinked and thermal compound used.

TESTING THE SCR

SCRs can be tested with an ohmmeter by connecting the positive lead to the anode and the negative lead to the cathode, figure 17-9. When an ohmmeter is connected across the anode-cathode

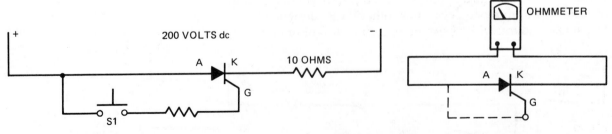

Figure 17-8 The gate will turn the SCR on but not off.

Figure 17-9 Ohmmeter test

circuit of the SCR, there is infinite or high resistance. This high resistance value is seen since the SCR is off unless triggered by the gate.

With the leads connected and using a jumper, touch the gate of the SCR to the positive ohmmeter lead. The SCR shows a great decrease of resistance. When the gate is disconnected from the positive lead, the SCR may turn off or continue to conduct depending on the device. If it has a small holding current, the ohmmeter may supply enough current to keep the SCR turned on. If the ohmmeter cannot supply enough holding current, the SCR will turn off.

It should be noted that many of the high-power SCRs used in industry have an internal resistor connected between the gate and cathode as shown in figure 17-10. The purpose of this resistor is to help keep the SCR from false triggering due to line interference. This resistance can be measured with an ohmmeter when testing the SCR and can cause a maintenance electrician who is not aware of the existence of this resistor to diagnose the device as being leaky between the gate and cathode.

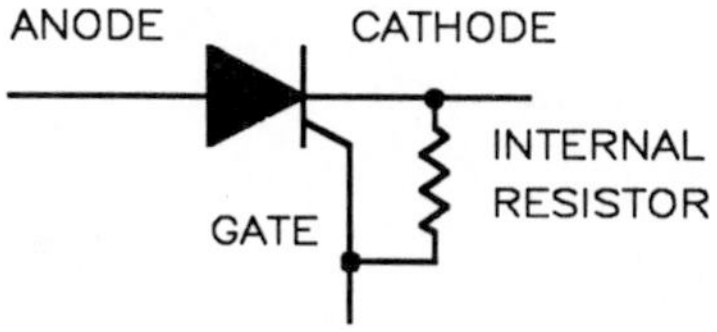

Figure 17-10 SCR with an internal resistor

APPLICATIONS

A good example of how an SCR can be used in a DC circuit can be seen in figure 17-11. This is a battery charger circuit that turns the power applied to the battery off when the battery reaches full charge. The transformer is a step-down transformer with a second-

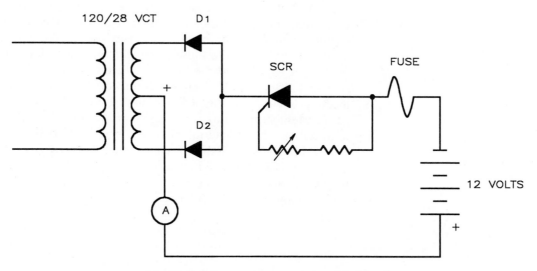

Figure 17-11 Battery charger with automatic turn-off

ary voltage of 28 volts. Notice that the secondary of the transformer has been center tapped. The two diodes, D_1 and D_2, form a full-wave rectifier with the center tap of the transformer forming the positive output lead. It is assumed this circuit is to be used to charge a 12 lead-acid battery.

When a 12 lead-acid battery reaches full charge, it will exhibit a terminal voltage of 14 volts. If the battery should be charged beyond this point, it will overcharge, which could damage the battery. To prevent overcharging, the battery should be charged to a voltage of less than 14 volts. The ideal voltage to charge a 12 lead-acid battery to is generally considered to be 13.8 volts.

In this circuit, the two-diode type of rectifier will produce a peak dc voltage of 19.8 volts ($14 \times 1.414 = 19.8$). When a battery is in a low-charge state, the terminal voltage is low, which permits the power supply to furnish current to the battery. As the state of charge of the battery increases, its terminal voltage will increase to a maximum of 14 volts. The amount of charging current will be much higher during the period of time the battery is in a low state of charge. As the state of charge of the battery increases, the amount of charging current will decrease to a very small amount.

The amount of gate current supplied to the SCR is determined by the amount of resistance connected in the gate circuit, and the difference in voltage between the power supply and the terminal voltage of the battery. When the battery is in a low-charge state, the difference in voltage is great enough to permit enough gate current to flow to turn the SCR on. Once the SCR has been turned on by the gate, it will remain on due to the relatively high-charging current. As the state of charge of the battery increases, two things happen at the same time. The charge current decreases, and the terminal voltage of the battery increases. As the terminal voltage of the battery increases, there is less difference in voltage between the power supply and the battery, which causes the gate current supplied to the SCR to decrease. If the variable resistor connected in the gate circuit of the SCR is adjusted properly, the amount of gate current supplied to the SCR will drop below the level needed to turn it on when the terminal voltage of the battery reaches about 13.8 volts. If the SCR has been selected correctly, the charging current at this point will have dropped below the holding level of the SCR. This permits the SCR to turn off and stop conduction of the charging current completely.

UNIT 18

The SCR in an AC Circuit

When connecting an SCR in an ac circuit, the first thing to remember is that the SCR is a rectifier and the output of the SCR is pulsating direct current. The only difference between an SCR and a common rectifier diode is that the SCR can be triggered to turn on at a specific point in the waveform.

DC AND AC CONNECTIONS

When the SCR is connected into a dc circuit, the gate can turn the SCR on, but cannot turn it off. The current through the anode-cathode circuit of the SCR must drop below the holding current level before the device will turn off.

The same is true for an SCR connected into an ac circuit. The gate still controls the turn on of the SCR. The anode-cathode current must drop below the holding current level for the device to turn off. The main difference in operation is caused by the ac power itself. The gate turns the SCR on, but the ac waveform dropping back to 0 volt will turn it off. It is therefore necessary to retrigger the gate of the SCR for each half cycle of alternating current conducted.

OPERATION OF THE CIRCUIT

Consider the circuit shown in figure 18-1. In this circuit, the gate of the SCR is connected through a resistor and diode directly to its anode. When the ac voltage applied to the anode rises in the positive direction, current flows through the gate-cathode section of the SCR. When it reaches the trigger point (assume 5 milliamperes (ma) for this SCR), the SCR fires and conducts through the anode-cathode section. It conducts as long as the ac voltage remains in the positive direction and the current is above its holding

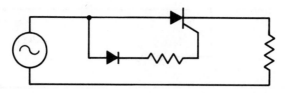

Figure 18-1 Basic SCR circuit

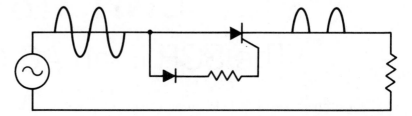

Figure 18-2 SCR with no control of the gate current

current level. When the ac voltage drops to zero and begins to increase in the negative direction, the SCR remains turned off. When the voltage applied to the anode again becomes positive the gate will trigger the SCR on again. The output of SCR is half-wave rectified direct current, figure 18-2.

Adding the Variable Resistor

Why use an SCR if it does the same thing as a simple junction diode? The SCR can be controlled as to *when* it turns on, figure 18-3. A variable resistor has been added to the gate circuit of the figure. Assume that the gate current must reach a 5 ma level before the SCR will fire. By adjusting the variable resistor, it is possible to determine how much voltage must be applied to the gate before a 5 ma current flows through the gate circuit. By adjusting the resistor to a higher value, it is possible to keep the SCR from firing until the ac voltage has reached its peak value in the positive direction. In this way the SCR will fire when the ac voltage is at its positive peak. With this setting of the resistor, the SCR will drop half the voltage and the load will drop the other half, figure 18-4.

REDUCING THE RESISTANCE: By reducing the resistance of the gate circuit, the gate current reaches 5 ma sooner, and the SCR fires earlier in the ac cycle. This causes less voltage to be dropped across the SCR and more to be dropped across the load, figure 18-5.

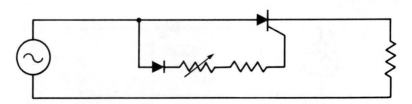

Figure 18-3 Variable resistor controls gate current

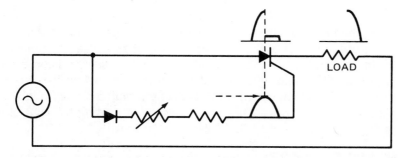

Figure 18-4 SCR fired at the peak of the waveform

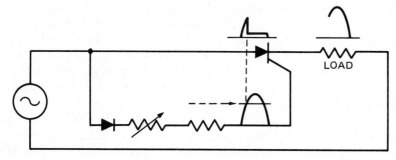

Figure 18-5 SCR fired before the waveform reaches peak

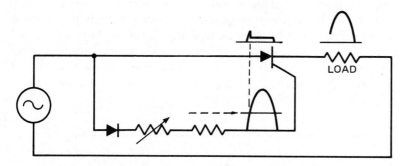

Figure 18-6 SCR fired earlier than in figure 18-5

If the gate resistance is reduced even more, the 5 ma of gate current will be reached even sooner during the cycle and the SCR will fire earlier. Still less voltage is dropped across the SCR and more voltage is dropped across the load, figure 18-6.

PROBLEMS MET ON THE WAY TO FULL CONTROL

There is a problem with this type of control, however. The SCR only controls half of the positive cycle of alternating current applied to it, figure 18-5. The latest that the SCR can be fired is when the positive half cycle has reached 90°. This permits the SCR to control only half of the ac positive wave; half voltage is applied to the load when the SCR initially fires. If the load resistor is a light bulb, the bulb will burn at half brightness when it is first turned on. The SCR circuit controls the light bulb from half brightness to full brightness. There are methods of gaining full control of the waveform; these will be discussed in unit 19.

APPLICATIONS

The circuit shown in figure 18-7 is an example of how an SCR can be used in an ac circuit. In this circuit, the SCR is used as a touch controller. When the touch plate is touched by a person, the SCR turns on and permits current to flow through the load. The SCR will remain turned on as long as the person continues to touch the plate. This circuit can be used when it is necessary to have a switch with no moving or mechanical contacts.

Notice that a neon lamp has been connected in series with the gate of the SCR. The neon lamp is used because it is a gas-filled tube and depends on ionization of the gas for operation. Most devices that operate by ionization have a characteristic known as negative resistance. This means that the device will start conduction or turn on at one voltage and continue conduction at a lower voltage. For example, assume that this lamp requires a potential of 15 volts across it to ionize or start conduction. Once the gas in the tube has ionized, it will continue to conduct at a much lower voltage, say 5 volts. As long as the voltage across the tube is

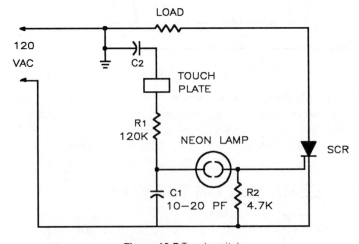

Figure 18-7 Touch switch

greater than 5 volts it will continue to conduct. When the voltage drops below 5 volts, the tube will stop conduction or turn off, figure 18-8.

The circuit operates due to the voltage divider formed by capacitors C_1 and C_2. Capacitor C_2 is actually the body capacitance of a person. The largest amount of voltage will be dropped across the capacitor with the lowest amount of capacitance. When no one is touching the touch plate, the circuit between the touch plate and ground is open. This forms a very low value of capacitance causing almost all the circuit voltage to be dropped across these two points. When the touch plate is touched, the capacitance of C_2 becomes much greater than that of C_1. This causes the voltage across capacitor C_1 to become higher than 15 volts, causing the neon lamp to ionize and start conduction. When the neon lamp ionizes, its internal impedance drops immediately, permitting capacitor C_1 to discharge and supply a pulse of current to the gate of the SCR. The SCR turns on and allows current to flow through the load.

Resistor R_1 limits current in the gate circuit, and resistor R_2 is used to ensure the SCR will remain turned off until the neon lamp ionizes and discharges capacitor C_1. The sensitivity of this circuit is determined by the area of the touch plate. If the area becomes great enough, this control will become a proximity detector and will trigger when a person comes near the touch plate without actually touching it.

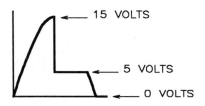

Figure 18-8 The voltage must drop below 5 volts before conduction will stop.

UNIT 19

Phase Shifting an SCR

OBJECTIVES

After studying this unit the student should be able to:

- Discuss the meaning of phase shifting an SCR
- Discuss the reasons for phase shifting an SCR
- Construct a circuit for phase shifting an SCR in an ac circuit

In order for an SCR to gain complete control of the waveform it must be *phase shifted*. This means to change or shift the phase of one thing in reference to another. In this case, the concern is with shifting the phase of the voltage applied to the gate with respect to the voltage applied to the anode.

PRACTICAL APPLICATION

Recall that an SCR which has not been phase shifted can control only half of the positive waveform. An SCR which has not been phase shifted has the voltage which is applied to the gate locked in phase with the voltage applied to the anode, figure 19-1. Assume that the SCR in this example fires when the gate current reaches a 5 ma level.

The variable resistor in figure 19-2 is adjusted so that the gate current will not reach the 5 ma level until the applied voltage reaches its peak value. The SCR fires when the voltage applied to

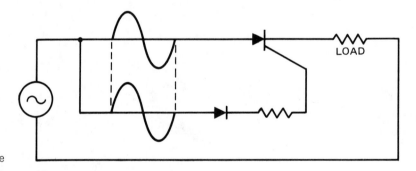

Figure 19-1 Voltage applied to the gate in phase with the voltage applied to the anode

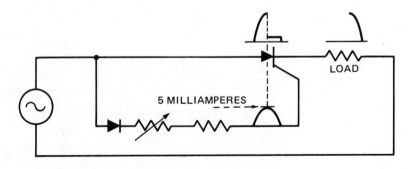

Figure 19-2 SCR fired at the peak of the waveform

94

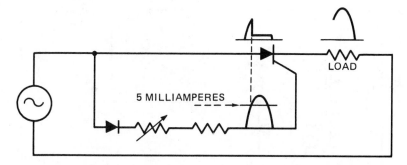

Figure 19-3 SCR fired before the waveform reaches the peak

the anode is at the peak level also. If the variable resistor is decreased in value, the gate current can reach 5 ma sooner, and the SCR will fire sooner during the cycle, figure 19-3.

Each decrease of the variable resistor will cause the gate current to reach 5 ma earlier in the cycle and cause the SCR to fire sooner, figure 19-4.

Separating Voltage

Notice that the gate controls the firing of the SCR during only half of the positive cycle. As long as the voltage applied to the gate of the SCR is locked in phase with the voltage applied to the anode, the SCR cannot be fired after the ac voltage applied to the anode has reached 90°. To obtain full control of the ac half cycle applied to the anode, the voltage applied to the gate must be out-of-phase with the voltage applied to the anode. To shift the phase of the voltage applied to the gate, the gate voltage must be separated from the voltage applied to the anode. There are several ways to do this, but for now only one of these methods will be investigated.

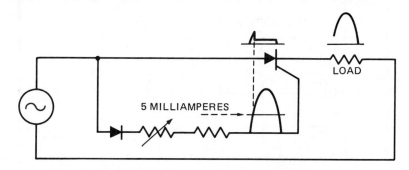

Figure 19-4 SCR fired very early in the waveform

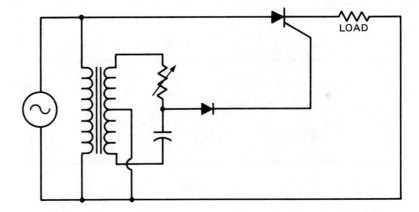

Figure 19-5 Isolation transformer used to phase shift SCR

CAPACITIVE PHASE SHIFT: One of the simplest ways to phase shift an SCR is to use a small center-tapped transformer to provide trigger circuit isolation from the line, figure 19-5. The variable resistor and capacitor in this circuit causes the gate current to be capacitive in comparison to the anode-cathode current. Since the current in a capacitive circuit leads the voltage, the gate current is shifted out-of-phase with the voltage. The variable resistor determines the amount of phase shifting. By shifting the gate current out-of-phase with the voltage, the gate current reaches 5 ma after the applied voltage reaches its peak value, figure 19-6. Since the gate current reaches the 5 ma level after the voltage applied to the anode reaches its peak value, the SCR fires later in the cycle. This causes a higher voltage to be dropped across the SCR and less voltage dropped across the load.

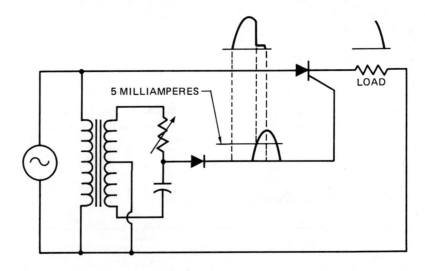

Figure 19-6 SCR fired after the waveform has passed the peak

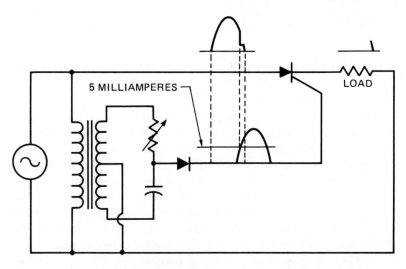

Figure 19-7 SCR fired late in the waveform

If the variable resistor is adjusted so that the gate current leads the voltage by a greater amount, the SCR fires later, figure 19-7. The SCR therefore drops a greater amount of voltage and the load has less voltage applied to it. By choosing the correct values of resistance and capacitance, the voltage applied to the load can be controlled from zero to its full value.

UNIT 20
UJT Phase Shifting for an SCR

OBJECTIVES

After studying this unit the student should be able to:

- Discuss the use of a UJT as a phase shifting device for an SCR
- Discuss the operation of a Shockley diode
- Draw the schematic symbol of a Shockley diode
- Discuss the use of a Shockley diode as a phase shifting device for an SCR
- Discuss the characteristics of a silicon unilateral switch
- Draw the schematic symbol of a SUS
- Construct a phase shift circuit for an SCR using a UJT

The unijunction transistor (UJT) was developed to do the job of phase shifting SCR circuits. It is an ideal device for this application since its operating characteristic is that of a pulse timer.

An SCR must be phase shifted to gain complete control of the ac waveform supplied to it. The UJT offers an easy way of accomplishing this job, figure 20-1. This circuit can be broken down into two separate sections. One is the power handling part and consists of the SCR and the load. The other is the brains and consists of the step-down transformer, UJT, and related components.

The transformer in this circuit is used to provide a low voltage to operate the UJT. The UJT is turned off until capacitor C1 charges to a predetermined voltage. Assume this voltage to be 10 volts. The charge time for the capacitor is determined by its capacitance and the resistance of resistor R1. When the charge on the capacitor reaches a value of 10 volts, the UJT turns on and discharges the capacitor through resistor R2. This discharge produces a pulse of current at R2 which triggers the gate of the SCR. When the voltage across C1 drops to about 3 volts, the UJT turns off and permits the capacitor to begin charging again. Since the charge time of the capacitor is determined by the resistance of R1, the pulse rate of the UJT is controlled by varying the resistance of resistor R1.

The pulses produced by the UJT are entirely independent of the ac voltage connected to the anode of the SCR. Since the UJT can be triggered at any time regardless of the ac waveform, the SCR

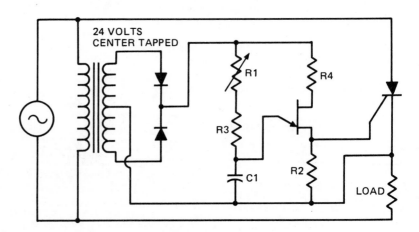

Figure 20-1 Unijunction transformer (UJT) phase-shift circuit

can be fired at any point during the positive half cycle of alternating current applied to it.

Phase shifting SCR circuits with the unijunction transistor has become a very common practice in industrial electronic controls. It is important to gain a good understanding of this and other phase-shifting circuits.

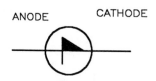

Figure 20-2 Shockley diode

OTHER TRIGGERING DEVICES

Although the unijunction transistor is one of the most common devices used for phase shifting SCR circuits, it is not the only device that can be used. Two other devices that can be implemented to phase shift an SCR are the four-layer diode or *Shockley diode* and the *SUS* or Silicon Unilateral Switch.

The Shockley diode has a negative resistance characteristic very similar to the UJT. The diode will remain turned off until its break-over voltage is reached. At this point the diode turns on and conducts until the voltage across the device drops to its turn off value. The break-over or turn on value of voltage is higher than the turn off value. For example, its turn on value may be 8 volts and the turn off value may be 3 volts. The device's ability to conduct current at a lower voltage value than it takes to turn it on makes it useful for phase shifting. The schematic symbol for a Shockley diode is shown in figure 20-2.

A simple SCR phase shifting circuit using the Shockley diode is shown in figure 20-3. When the ac waveform applied to the anode of diode D_1 becomes positive, the diode turns on and begins to conduct current through resistors R_1 and R_2. This permits capacitor C_1 to begin charging. When the voltage across capacitor C_1 increases to the break-over voltage of the Shockley diode, it turns on and discharges capacitor C_1 through the gate of the SCR. This

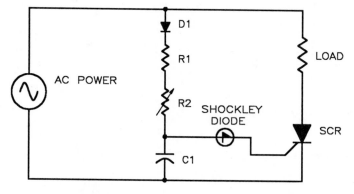

Figure 20-3 Shockley diode used to phase shift an SCR

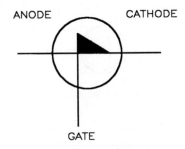

ANODE   CATHODE

GATE

Figure 20-4 Silicon unilateral switch

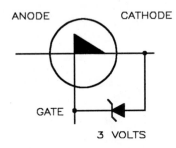

ANODE   CATHODE

GATE

3 VOLTS

Figure 20-5 Zener diode used to adjust the break-over voltage of an SUS

pulse of gate current causes the SCR to turn on and conduct current through the load. When C_1 has discharged to a voltage that is below the turn off value of the Shockley diode, it turns off. The next positive half cycle of ac voltage begins charging the capacitor again and the sequence of events is repeated.

The amount of time necessary for capacitor C_1 to reach the break-over value of voltage for the Shockley diode is determined by the value of capacitance of capacitor C_1 and the sum of the resistances of resistors R_1 and R_2. Resistor R_2 has been made variable to permit the charge time of capacitor C_1 to be adjusted. If the values of capacitor C_1 and resistor R_2 are chosen correctly, the SCR can be phase shifted over the entire half cycle.

The silicon unilateral switch is very similar to the Shockley diode. The basic difference between the two devices is that the SUS has a gate lead which permits some control of the value of break-over voltage needed to turn the device on. The schematic symbol of the SUS is shown in figure 20-4. The break-over voltage value can be adjusted by connecting a zener diode between the gate and cathode as shown in figure 20-5. In this example, a 3-volt zener diode is used. This permits the break-over voltage to become about 3.6 volts (3 volts of the zener diode + 0.6 volts needed to turn on any silicon device). The break-over voltage can be adjusted to a value that is less than the original value if no zener diode were to be used, but it cannot be adjusted to a higher value.

UNIT 21

SCR Control of a Full-wave Rectifier

Up to this point, the control of only one SCR in a circuit has been studied. When only one SCR is used, the output voltage is half-wave direct current. Almost all industrial applications for SCRs require full-wave rectification of the ac voltage. SCRs can control the output of a bridge rectifier when connected into a bridge circuit as shown in figure 21-1.

In this figure, two SCRs have been connected so that their cathodes are tied together to form the positive post of the bridge rectifier. The other two rectifiers are common junction diodes. Since each half of the ac waveform must pass through one of the SCRs, only two are needed for full control of the bridge rectifier.

The phase-shift network can be the same as for a single SCR, figure 21-2. This circuit can be separated into two sections. One section is the bridge containing the SCRs and the load resistor. The other is the phase-shifting network made up of the step-down transformer, UJT, and related components. This UJT circuit is identical to the circuit used to provide phase-shift control for a single SCR. The only real difference is that this circuit has the gate leads of each SCR tied together through a low value of resistance. These resistors make the gates of the SCRs more closely matched in impedance; one SCR will not fire ahead of the other and take all the gate current. The two low-value resistors force the gates to share the current delivered by the UJT.

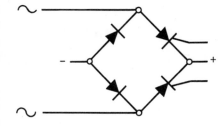

Figure 21-1 Bridge rectifier controlled by SCRs

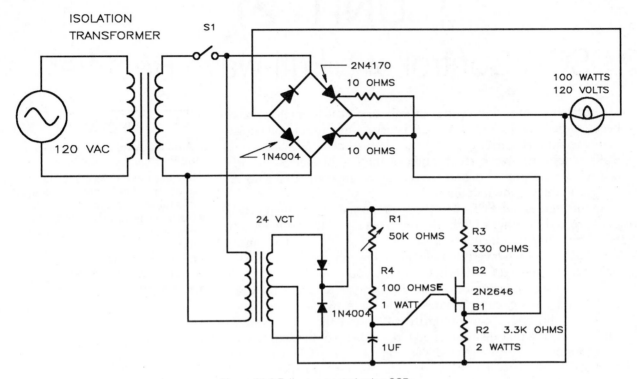

Figure 21-2 Full-wave control using SCRs

UNIT 22

A Solid-state Alarm with Battery Backup

The alarm circuit in figure 22-1 operates on 120 volts ac (120-Vac), during normal operation. If the ac power is interrupted, a 12-volt battery is used to provide continued operation of the circuit. Battery operation continues until the ac power is restored. At this time the circuit returns to operation on the ac power line. Sensor switches used on doors and windows are normally closed and form one continuous series circuit.

OBJECTIVES

After studying this unit the student should be able to:
- Discuss the operation of this alarm system
- Construct this circuit using discrete components

PARTS OF THE ALARM

The parts of this alarm include:

- Fuse F1 is a 1-amp buss-type fuse. It is used to protect the circuit against a short circuit.
- Switch S1 is a double-pole switch. One side of it is used to control the ac voltage to the primary of the transformer. The other side disconnects the 12-volt battery from the dc side of

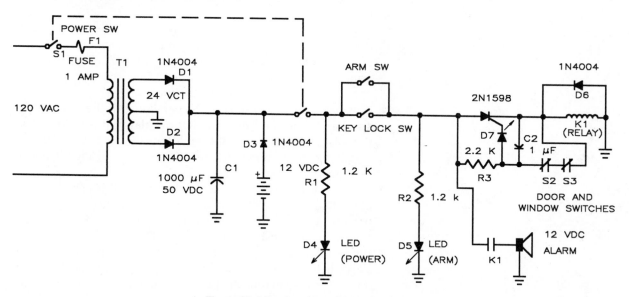

Figure 22-1 Burglar alarm with battery backup

the circuit. The dotted line between the two switches shows mechanical connection. It is the power switch and connects power to the circuit.

- T1 is a 24-volt center-tapped transformer used to step the 120-Vac down to 24-Vac.
- Diodes D1 and D2 are used as rectifiers to change the 24-Vac into dc. Since this is a center-tapped transformer, the dc voltage is about 12 volts.
- Capacitor C1 is used to filter the output voltage of the rectifier. When the ripple has been filtered, the average dc voltage increases to about 17 volts.
- Diode D3 is connected in series with the 12-volt battery which supplies the backup power source. Diode D3 prevents current from being drawn from the battery unless the ac power is interrupted. When the rectifier is operating, 17-volts dc (17-Vdc), is applied to the cathode of diode D3 and 12 volts is applied to its anode. In this condition, diode D3 is reverse biased and no current flows. If the ac power is interrupted, however, the output of the rectifier turns off. When the voltage of the rectifier drops below 12 volts, diode D3 becomes forward biased and the battery supplies the power for the continued operation of the circuit.
- Diodes D4 and D5 are light-emitting diodes used as pilot lights. Diode D4 indicates that the power switch has been turned on. Diode D5 indicates that the alarm has been armed.
- Resistors R1 and R2 are used to limit the current to D4 and D5.
- The key switch is a key-locked switch installed at an outer door of the building the alarm is to protect. Upon leaving the building the key switch is closed and the alarm becomes armed. Upon returning to the building the key switch is opened, disarming the alarm circuit.
- The arm switch is connected in parallel with the key switch, and located on the front panel of the alarm. It is used to arm the system from inside the building. (Note: If the alarm is installed inside a home, it can be armed from inside the house at night.)
- The SCR is the real controller of the circuit. It is used because of its characteristic of remaining turned on once the gate has been triggered. When the SCR turns on, it provides the current to turn on relay coil K1.
- When relay K1 turns on, its contacts close and connect the alarm bell or siren to the power line. With the SCR turned on, the arm switch or the power switch must be reopened to turn the alarm off.

- Resistor R3 is used to limit the gate current of the SCR.
- Capacitor C2 helps prevent false triggering of the SCR due to voltage spikes which may appear on the power line.
- Diode D6 is used as a kick-back or free-wheeling diode to kill the voltage spike induced in the coil of relay K1 when the power is interrupted.
- Switches S2 and S3 are normally closed switches connected to doors, windows, or whatever is to be protected. The schematic shows two of these switches, but any number can be used as long as they are normally closed and connected in series. As long as these switches remain closed, the gate of the SCR is connected to ground and cannot be triggered. If one of them opens, however, it permits current to flow to the gate and trigger the SCR.

OPERATING THE ALARM

The operation of the circuit is as follows:

1. When switch S1 is closed, dc voltage is provided by the rectifier. This permits diode D4 to light to indicate that the power switch is turned on.
2. With either the key switch or the arm switch closed, power is supplied to the SCR. Diode D5 lights, indicating that the circuit is armed.
3. Since the door and window switches are closed, the gate current of the SCR is connected to the cathode and the SCR will not trigger. If one of them is opened, however, current is provided to trigger the gate of the SCR.
4. When the SCR turns on, relay K1 turns on and closes the contacts which provide power to the bell or siren. Once the SCR has turned on, it cannot be turned off unless the arm switch or power switch is opened.

If the ac power should fail after the power and arm switch have been turned on, the battery will provide the power for the operation of the circuit. When the ac power is restored, the battery is disconnected from the circuit. This circuit does not require the use of a large battery since it is only used when the ac power is interrupted. A 12-volt lantern battery, which can be purchased at most sporting goods stores, is ideal for this circuit.

UNIT 23

The Diac and Silicon Bilateral Switch

The diac is a special-purpose bidirectional diode. Its primary function is to phase shift a triac. The trigger operation of the diac is similar to that of a unijunction transistor, except the diac is a bi- or *two-directional device*. It can operate in an ac circuit while the UJT is a dc device only.

There are two schematic symbols for the diac, figure 23-1. Either symbol is used in an electronic schematic to illustrate the use of a diac, so become familiar with both.

The diac is a voltage-sensitive switch which can operate on either polarity, figure 23-2. When voltage is applied to the diac, it remains in the turned-off state until the applied voltage reaches a predetermined level. Assume this to be 15 volts. Upon reaching 15 volts, the diac turns on (fires). When the diac fires, it displays a *negative resistance*, meaning that it conducts at a lower voltage than that which was applied to it (assume 5 volts). The diac remains turned on until the applied voltage drops below its conduction level, or 5 volts, figure 23-3.

Figure 23-1 Schematic symbols of a diac

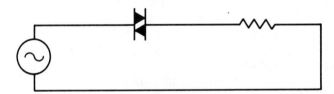

Figure 23-2 Diac operates on either polarity

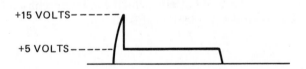

Figure 23-3 The diac conducts at a lower voltage than is required to turn it on.

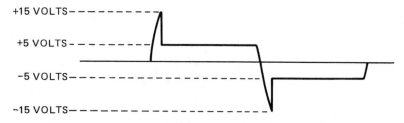

Figure 23-4 The diac conducts both halves of the ac waveform

Since the diac is a bidirectional device, it conducts on either half cycle of the ac voltage applied to it, figure 23-4. It has the same operating characteristic with either half cycle of alternating current. The simplest way to sum up the operation of the diac is to say it is a voltage-sensitive ac switch.

THE SILICON BILATERAL SWITCH

The silicon bilateral switch or *SBS* is another bidirectional device often used to trigger the gate of a triac. The schematic symbol for an SBS is shown in figure 23-5. The SBS is very similar to the diac in many ways. Both will conduct current in either direction, and both exhibit negative resistance. The SBS, however, has a more pronounced negative resistance region than does the diac. A characteristic voltage curve for both the diac and the SBS are shown in figure 23-6. Notice that the break-back voltage of the SBS is much more pronounced than that of the diac. Also, the break-over voltage of an SBS is generally lower than that of a diac. The break-over voltage of a diac will most often range between

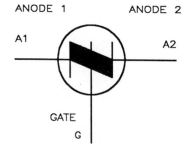

Figure 23-5 Silicon bilateral switch

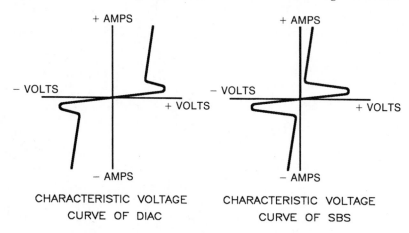

Figure 23-6 Characteristic voltage curve of diac and SBS

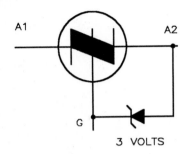

Figure 23-7 Zener diode used to control break-over voltage

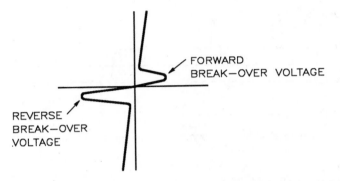

Figure 23-8 Forward and reverse break-over voltages are different.

± 16 volts and ± 32 volts. The most common break-over voltage for a silicon bilateral switch is ± 8 volts.

The silicon bilateral switch has several other advantages over the diac. SBSs are generally symmetrical to within about 0.3 volt. This means that the positive break-over voltage and the negative break-over voltage will be within about 0.3 volt of each other. The diac is generally symmetrical to within about 1 volt.

The SBS also has a gate lead that permits some control over the break-over voltage. If a 3-volt zener diode were to be connected between the gate and anode 2 as shown in figure 23-7, the positive break-over voltage would be reduced to about 3.6 volts (3 volts for the zener diode + 0.6 volts needed to turn on a silicon device), but the reverse break-over voltage would be unaffected as shown in figure 23-8. This could be used if it were desirable to have different forward and reverse break-over values, which is not usually the case. The SBS is most often used with the gate lead not connected.

UNIT 24
The Triac

The triac is a device very similar in operation to the SCR. Even their appearance is similar, as shown in figure 24-1, which shows triacs in various case styles. When an SCR is connected into an ac circuit, the output voltage is rectified to direct current. The triac, however, is designed to conduct on both halves of the ac waveform. Therefore, the output of the triac is alternating current instead of direct current.

The triac is made like two SCRs connected in parallel facing in opposite directions with their gate leads tied together, figure 24-2.

OPERATION OF THE TRIAC

Notice that the schematic symbol for the triac, figure 24-3, is similar to the connection of the two SCRs. The gate must be connected to the same polarity as MT2 to turn the triac on. When the voltage applied to MT2 increases in the positive direction, the gate fires the half of the triac which is forward biased during that half of the cycle. Since the other SCR half of the triac is reverse biased during that half cycle, it cannot be fired. When the applied ac

Figure 24-1 Triacs shown in various sizes and case styles

PHOTO COURTESY OF GENERAL ELECTRIC CO.

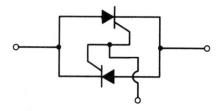

Figure 24-2 The triac operates like two SCRs connected together.

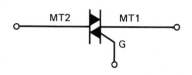

Figure 24-3 Schematic symbol of a triac

109

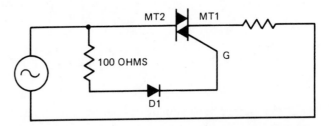

Figure 24-4 The diode permits only one half of the triac to conduct.

voltage becomes negative at MT2, the gate fires the other half of the triac which is now forward biased. This action can be seen with an oscilloscope, figure 24-4.

Adding a Diode

If a diode is connected in series with the gate, it permits only half of the ac cycle to provide gate current. The gate provides current flow for only half of the ac cycle: assume the positive half cycle. If an oscilloscope is connected across the load resistor, only the positive half of the ac waveform is seen. The diode is reverse biased during the negative half cycle and no gate current flows to turn the negative half of the triac on.

If the diode is removed from the circuit and replaced facing in the opposite direction, it blocks the positive half cycle and passes the negative half. The oscilloscope connected across the load resistor shows the negative half of the ac waveform. This is a good example of the two SCR arrangement of the triac.

Adding a Variable Resistor

If a variable resistor is used to control the gate of the triac, the same control problems exist as did with the SCR. A simple resistor control, as shown in figure 24-5, permits the triac to control only half of the waveform just like an SCR which has not been phase shifted.

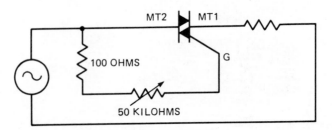

Figure 24-5 Gate current controlled by a variable resistor

TESTING THE TRIAC

The triac can be tested with an ohmmeter; since it is basically two SCRs, figure 24-2, it can be tested the same way. To test the triac, connect the meter leads of the ohmmeter to MT1 and MT2 of the triac. The meter should indicate no continuity until the gate lead is touched to MT2. One of the SCRs in the triac has now been tested. To test the other, reverse the meter leads connected to MT1 and MT2. No continuity should be read until the gate lead is again touched to MT2. By reversing the polarity of the ohmmeter, both of the SCRs in the triac can be tested.

APPLICATIONS

The ability of a triac to control a large amount of load current with a small amount of gate current can be seen in the circuit shown in figure 24-6. In this circuit, a small 24-volt thermostat is used to control the operation of a fan motor. It is assumed that the maximum current rating of the thermostat contacts is 0.5 amp, and the full load running current of the motor is 7.5 amps. If the thermostat contacts were to be used to control the operation of the motor, they would be destroyed the first time the motor was turned on.

If a triac is used as the controlling element for the motor, however, the thermostat can easily furnish the amount of gate current necessary to control the triac. A 24-volt step-down transformer is used to reduce the voltage in the thermostat circuit. The resistor limits the amount of current flow in the gate-MT1 circuit when the thermostat contacts close. Notice that one side of the secondary of the transformer is connected directly to MT1. This must be done to provide a complete path for current flow when the thermostat contacts close.

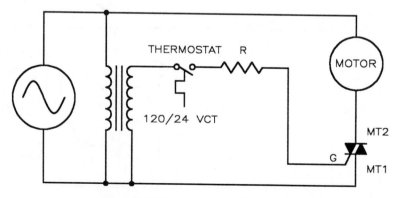

Figure 24-6 Triac controls motor operation

UNIT 25

Phase Shifting the Triac

OBJECTIVES

After studying this unit the student should be able to:

- Discuss phase shift control for a triac
- Construct a triac circuit using discrete components
- Make electrical measurements using test equipment

The triac, like the SCR, must be phase shifted if complete control is to be gained over the ac voltage applied to it. To phase shift the triac, separate the gate pulses from the voltage applied to MT1 and MT2. This is the same method used with the SCR.

One method of phase shifting the SCR is to use the unijunction transistor to supply pulses to the gate. Since the UJT is a dc operated device, however, it cannot be used to supply the trigger pulses for the triac which needs both negative and positive pulses. The device most often used to phase shift the triac is the diac, which is a bipolar device. It can supply both the negative and positive pulses needed to trigger the triac, figure 25-1.

Since the diac operates like an ac voltage-sensitive switch, assume that it turns on when the voltage applied to it reaches 30 volts and off when the voltage drops to 5 volts. In the operation of the circuit in figure 25-1, capacitor C1 is charged through resistors R1 and R2. The diac remains off until the voltage of capacitor C1 reaches 30 volts. When C1 reaches this value, the diac turns on and discharges C1 through the gate of the triac. This discharge pulse fires the triac. When the charge on the capacitor drops to 5 volts, the diac turns off again.

The phase-shift circuit in figure 25-1 is connected in parallel with the triac. Since the triac is essentially two SCRs, when it fires in one direction the voltage drop across the device becomes about 1 volt. It remains there until the ac wave drops back to zero and turns the device off. Once the diac fires and discharges capacitor

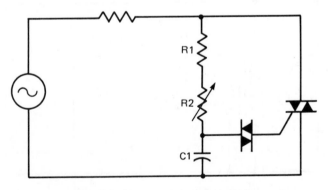

Figure 25-1 Diac used to phase shift a triac

C1, the capacitor cannot begin to charge again until the triac has turned off at the end of the ac cycle. As long as the triac is conducting there is only about 1 volt applied to the phase-shift circuit.

When capacitor C1 again begins to charge, it will charge at the opposite polarity from the preceding cycle. The diac turns on when the voltage across C1 reaches 30 volts and discharges the capacitor through the gate of the triac. When the triac fires, it conducts the other half of the ac waveform. The voltage applied to the phase-shift circuit drops to about 1 volt as long as the triac is conducting. The capacitor can not charge until the triac turns off at the end of that half cycle of alternating current. The triac can be triggered by either half cycle of the ac voltage applied to it.

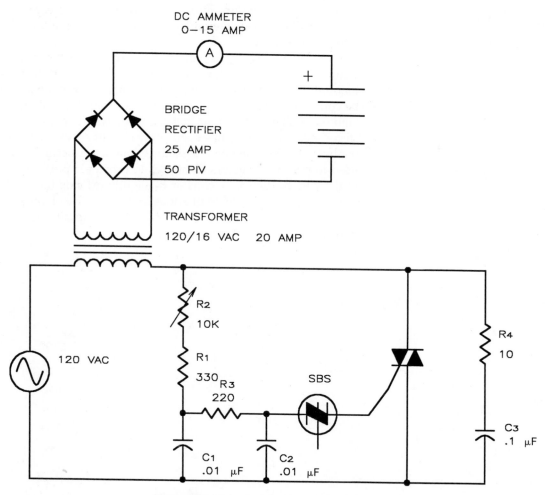

Figure 25-2 Triac-controlled battery charging circuit

The charge time of the capacitor is determined by the size of C1 and the amount of resistance connected in series with it. By varying the resistance of R2, the time it takes C1 to charge to 30 volts can be adjusted to different lengths. This allows the triac to be fired at different points along the cycle of ac voltage connected to it, which is the requirement for phase shifting a thyristor.

APPLICATIONS

A good example of how a phase shifted triac can be used is seen in figure 25-2. In this circuit, a triac controls the amount of charging current to a battery. The triac controls the current flow through the primary winding of a transformer, which in turn controls the transformer's secondary current. A bridge rectifier converts the alternating current delivered by the secondary of the transformer into direct current.

Components R_1, R_2, R_3, C_1, and C_2 are part of the phase shift network for the triac. Resistor R_2 is variable, which permits the amount of current flow through the primary of the transformer to be controlled. The value of resistor R_1 sets an upper limit on the amount of charging current that can be delivered to the battery. A silicon bilateral switch has been connected in series with the gate of the triac. The negative resistance characteristic of the SBS is used to permit capacitors C_1 and C_2 to provide a pulse of current to the gate of the triac. Resistor R_3 slightly extends the discharge rate of capacitor C_1, which provides a current pulse of longer duration to the gate of the triac. Resistor R_4 and capacitor C_3 provide transient or spike voltage protection for the circuit.

UNIT 26
Other Methods of AC Voltage Control

An ac voltage can be controlled by electronic devices, the most common of which is the triac. There are other devices that can be used to control an ac voltage when properly connected; the SCR, for example. It can be used as an ac switch when connected in a circuit as shown in figure 26-1.

THE SCR FOR VOLTAGE CONTROL

In this circuit, a bridge rectifier is connected in series with the load resistor. If current is to flow to the load, it must pass through the SCR which is connected across the positive and negative points of the bridge rectifier.

Operating the Circuit with the SCR

In order to understand the operation of this circuit:

1. Assume that point X of the ac line is positive and point Y is negative, also that current flow is from positive to negative.
2. Current flows to point A of the rectifier.
3. Diode D1 is forward biased, so current flows to point B of the rectifier.

OBJECTIVES

After studying this unit the student should be able to:

- Discuss different methods of controlling an ac voltage
- Construct a circuit for controlling ac voltage using a bridge rectifier and an SCR circuit
- Construct a circuit for controlling ac voltage using a transistor and a bridge rectifier

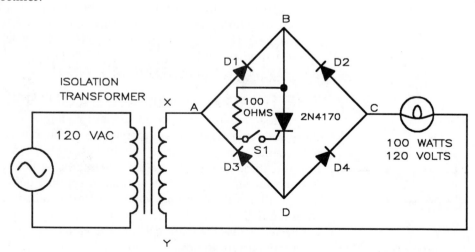

Figure 26-1 SCR ac switch

4. At point B, diode D2 is reverse biased, so current flows through the SCR to point D of the rectifier.
5. At point D, both diodes D3 and D4 are forward biased, and current does not flow from positive to positive.
6. The current flows through diode D4 to point C of the rectifier.
7. From here, the current flows through the load resistor to the other side of the ac line at point Y.
8. Since current flowed through the load resistor during the half cycle, a positive voltage appears across the load resistor.

Now, assume that point Y of the ac line has become positive and point X is now negative.

1. Current flows from point Y, through the load resistor to point C of the bridge.
2. The current flows through diode D2 to point B.
3. From point B, current flows through the SCR to point D of the rectifier.
4. Both diodes D3 and D4 are forward biased, so the current flows to the most negative point, or through diode D3 to point X of the ac line.
5. Since current flowed through the load resistor during the half cycle, a voltage appears across the resistor.
6. The current, however, flowed through the load resistor in the opposite direction. Therefore, a negative voltage appears across the resistor.

Notice that the current flowed through the load resistor in both directions, but it flowed through the SCR in only one direction. This type of circuit permits a dc device to control alternating current.

THE TRIAC FOR VOLTAGE CONTROL

The triac is most often used to control alternating current, but there are some conditions under which a triac can have undesirable characteristics. The triac operates like two SCRs connected in opposite directions, figure 26-2. If the internal structure of the triac is unbalanced, it causes one section to conduct before the other. The positive half of the ac wave will start conducting before the negative half. Figure 26-3 shows an oscilloscope which has been connected across the load resistor. The variable resistor is adjusted to permit current to flow through the load resistor. The half of the triac which controls the positive half of the ac wave begins to

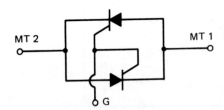

Figure 26-2 A triac operates like two SCRs connected together.

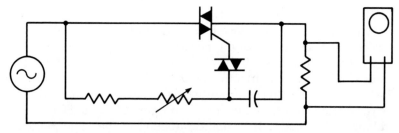

Figure 26-3 Triac control of ac voltage

conduct before the half that controls the negative portion. A waveform similar to the one shown in figure 26-4 is seen on the display of the oscilloscope. The waveform in figure 26-4 shows that only a portion of the positive half cycle is being conducted. The load resistor, therefore, has dc voltage applied to it.

LIMITATIONS TO CONTROL OF THE CIRCUIT

If dc voltage is applied to a pure resistance, no adverse conditions will occur. If dc voltage is applied to a load which is inductive, however, a great deal of damage can be done. When an ac voltage is applied to an inductive load, the current is limited by inductive reactance; with dc voltage, however, the current is limited only by the wire resistance of the coil. Most ac inductive loads such as motors or transformers have a low wire resistance. By applying dc voltage to one of these loads, the motor or transformer can be destroyed.

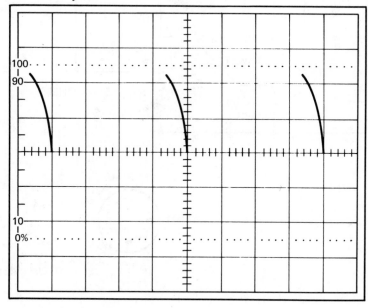

Figure 26-4 Only the positive half of the ac waveform is being conducted.

Solving the Problem

When an electronic device is used to control an ac voltage, the device must conduct both the negative and positive halves of the ac waveform, especially if the ac voltage is applied to an inductive load. If a phase shift circuit is added to the SCR shown in figure 26-1, both the negative and positive halves of the ac waveform will be conducted. In the circuit shown in figure 26-5, the SCR is controlling a dc voltage. It begins conducting at the same point for each dc pulse applied to it. Since one of the dc pulses permits

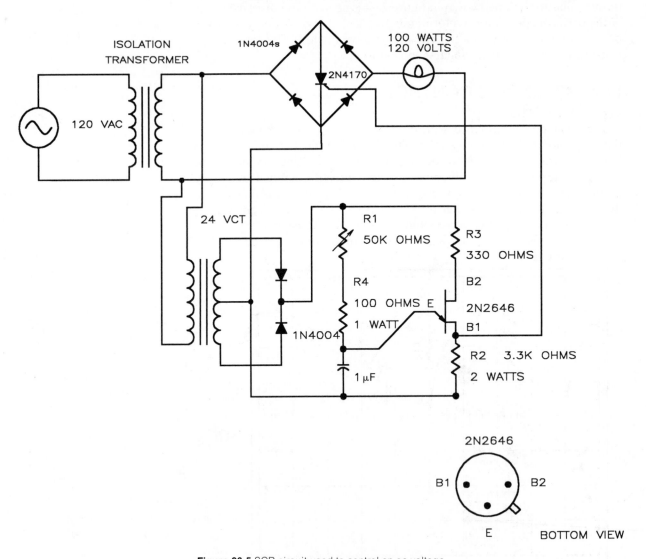

Figure 26-5 SCR circuit used to control an ac voltage

current to flow through the load resistor in one direction, and the next pulse permits current to flow through the load resistor in the opposite direction, the SCR permits both the positive and negative halves of the ac waveform to be conducted.

If an oscilloscope is connected to the load resistor in the circuit shown in figure 26-5, a waveform similar to the one shown in figure 26-6 can be seen. Notice that both the positive and negative halves of the ac waveform have been conducted.

The ac output voltage of the circuit in figure 26-5 is controlled by permitting the SCR to fire at different points in a cycle. The voltage applied to the load is determined by the amount of time the SCR is conducting compared to the amount of time it is not.

OTHER DEVICES USED FOR CONTROL

Thryistor devices such as the SCR and triac control the voltage by chopping the waveform. Resistive loads or motors operate without problems when this type of waveform is applied to them. Other devices, however, will not. If a variable ac voltage is applied to these devices, it must be done by increasing or decreasing the amplitude of the waveform. This is accomplished by replacing the SCR with a transistor, figure 26-7. The transistor operates by permitting more or less current to flow through it, not by turning

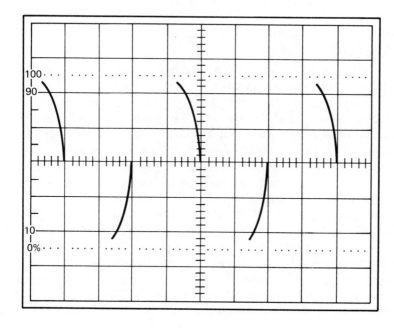

Figure 26-6 Both halves of the ac waveform are being conducted.

COPYRIGHT © 1983, TEKTRONIX, INC.
REPRODUCED BY PERMISSION

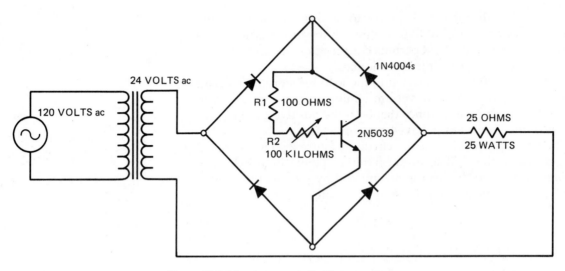

Figure 26-7 AC voltage controlled by a transistor

on or off at different points in the waveform. Since the transistor controls the ac voltage applied to the load by controlling the current flow, voltage is dropped across the collector-emitter of the transistor. Care must be taken not to exceed the transistor's voltage, current, or power ratings.

APPLICATIONS

As previously stated, when variable ac voltage is to be connected to an induction motor to perform the job of speed control, it is very important that both halves of the ac waveform are conducted to the stator of the motor. The circuit shown in figure 26-8 can be used to provide variable voltage to the windings of an ac induction motor. A bridge rectifier has been connected in series with the motor windings. An SCR is used to control the current flow through the bridge rectifier. Gate current for the SCR is provided by a circuit consisting of a step-down transformer with a center-tapped secondary winding, diodes D_1 and D_2, resistors R_1 and R_2, capacitor C_1, and a Shockley diode.

The transformer is used to reduce the ac line voltage from 120 volts to 24 volts. Diodes D_1 and D_2 form a two-diode type of bridge rectifier to provide full-wave rectification for the gate of the SCR. Resistor R_1 limits current to the gate of the SCR if the value of resistor R_2 should be adjusted to zero. The adjustment of resistor

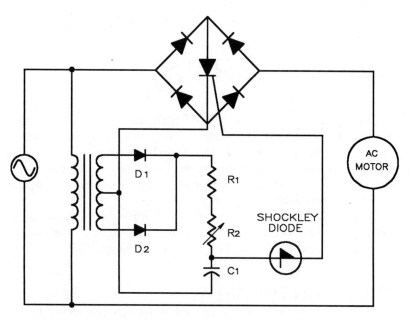

Figure 26-8 Variable voltage control for ac induction motor

R_2 determines at what point in the circuit the SCR will turn on. Capacitor C_1 serves two functions in this circuit. It produces a time delay by forming an RC time constant with resistors R_1 and R_2. It also provides a pulse of current to the gate of the SCR when the Shockley diode turns on and permits current to flow. The negative resistance characteristic of the Shockley diode is used to permit the SCR to be phase shifted.

UNIT 27
The Solid-state Relay

OBJECTIVES

After studying this unit the student should be able to:

- Discuss the operation of a solid-state relay
- Discuss different methods used to isolate the control section of the relay from the power section
- Connect a solid-state relay in an electrical circuit
- Discuss different uses and applications for solid-state relays

The solid-state relay is a device which has become increasingly popular for switching applications. The solid-state relay has no moving parts, is resistant to shock and vibration, and is sealed against dirt and moisture. Its greatest advantage however, is the fact that the control input voltage is isolated from the line device the relay is intended to control, figure 27-1.

Solid-state relays can be used to control either a dc or an ac load. If the relay is designed to control a dc load, a power transistor connects the load to the line as shown in figure 27-2.

THE LED IN THE RELAY

This relay has a light-emitting diode (LED) connected to the input or control voltage. When the input voltage turns the LED on, a photodetector connected to the base of the transistor turns the transistor on and connects the load to the line. This optical coupling is commonly used with solid-state relays. These relays are referred to as being *optoisolated*. This means that the load side of the relay is optically isolated from the control side of the relay.

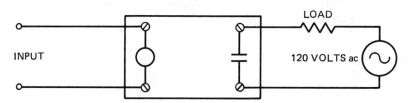

Figure 27-1 Solid-state relay

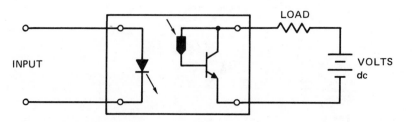

Figure 27-2 Solid-state relay used to control a dc load

122

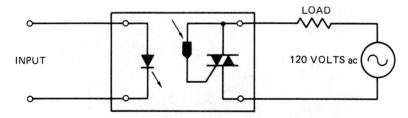

Figure 27-3 Solid-state relay used to control an ac load

The control medium is a light beam. No voltage spikes or electrical noise produced on the load side of the relay are therefore transmitted to the control side.

THE TRIAC IN THE RELAY

Solid-state relays intended for use as ac controllers have a triac connected to the load circuit in place of a power transistor. In figure 27-3, an LED is used as the control device just as in the previous example. When the photodetector sees the LED, it triggers the gate of the triac and connects the load to the line.

Other Devices in the Relay

Although optoisolation, figure 27-4, is probably the most common method used for the control of a solid-state relay, it is not the only method used. Some relays use a small reed relay to control the output, figure 27-5. A small set of reed contacts are connected to the gate of the triac. The control circuit is connected to the coil of the reed relay. When the control voltage causes a current to flow through the coil, a magnetic field is produced around the coil of the relay. This magnetic field closes the reed contacts and the triac turns on. In this type of solid-state relay a magnetic field, instead of a light beam, isolates the control circuit from the load circuit.

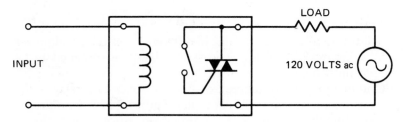

Figure 27-4 Reed relay used as a triac driver

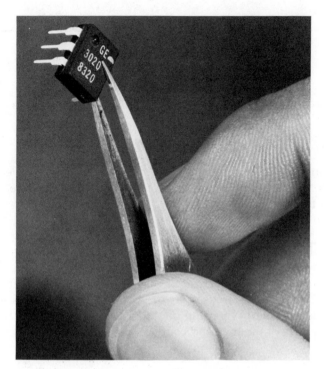

Figure 27-5 Control circuit isolated by a reed relay
PHOTO COURTESY OF GENERAL ELECTRIC CO.

Control Voltage

The control voltage for most solid-state relays ranges from about 3 to 32 volts and can be direct current or alternating current. If a triac is used as the control device, load voltage ratings of 120 to 240 volts ac are common. Current ratings can range from 5 to 25 amps. Many solid-state relays have a feature called zero switching. Assume that the ac voltage is at its positive peak value when the gate tells the triac to turn off. The triac continues to conduct until the ac voltage drops to a zero level before it actually turns off.

VARIOUS TYPES OF RELAYS

Solid-state relays are available in different case styles and power ratings, figure 27-6. Some solid-state relays are designed to be used as time-delay relays.

One of the most common uses for the solid-state relay is the I/O track of a programmable controller, figure 27-7. This controller is basically a small computer programmed to perform the same job as a control circuit using relays. Most of the actual control logic is

Figure 27-6 Different sizes and case styles for solid-state relays
PHOTO COURTESY OF MAGNECRAFT ELECTRIC CO.

performed by internal relays. These are digital-logic circuits programmed to operate like relays with normally-open or normally-closed contacts. Although the control logic is performed inside the computer, the computer must be able to communicate with the outside world to accomplish anything. The computer section of the programmable controller receives input information from sensing devices: float, pressure, and limit switches, as well as pushbuttons, figure 27-8. The computer is also able to give commands to the outside circuits to start and stop motors or open and close valves.

The computer section of the programmable controller operates on a dc voltage which generally ranges from 5 to 15 volts depending on the type of controller. The voltage source must be well

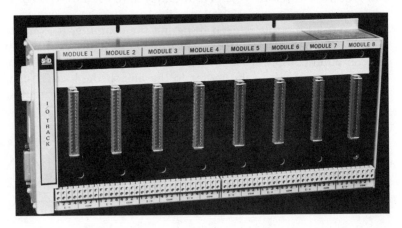

Figure 27-7 I/O track of programmable controller
PHOTO COURTESY OF STRUTHERS-DUNN, INC.

Figure 27-8 Processor unit of programmable control

PHOTO COURTESY OF STRUTHERS-DUNN, INC.

filtered, regulated, and free of voltage spikes and electrical noise. If a voltage spike reaches the computer, it can be interpreted as a command by the internal logic. For this reason there must be electrical isolation between the computer and the circuits outside it.

The function of the I/O track is to provide communication between the computer and the outside circuits while maintaining isolation between the two. Solid-state relays are used to perform this task. The I/O track contains both input and output modules.

An input module supplies information to the computer from an outside device. The circuit in figure 27-9 shows a float switch connected to an input module. When the switch closes, 120 volts ac is connected to the input of the solid-state relay. The relay provides a signal to the computer to tell it that the float switch has closed. This example shows 120 volts ac used as the input voltage. In actual practice, however, different types of programmable controllers require a different voltage. One unit may use low voltage ac while another may use low voltage dc.

The output module is used to permit the computer to communicate with the outside circuit. The circuits in figure 27-10 show the computer controlling the relay coil of a large motor starter. The motor starter connects the motor to the line. If the load is small enough, the solid-state relay is used as the controller without an

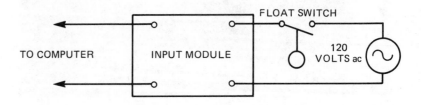

Figure 27-9 Float switch connected to an input module

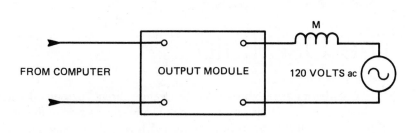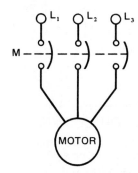

Figure 27-10 Output module controls relay coil

intermediate device. A solenoid valve, for example, could be controlled directly by the output module.

These examples show how solid-state relays can be used. The advantages of the solid-state relay are evident, and an increase in their use is expected.

UNIT 28

The Oscillator

OBJECTIVES

After studying this unit the student should be able to:

- Discuss the operation of an oscillator
- Discuss different uses for the oscillator
- Construct a square wave oscillator using discrete electronic components

Oscillators are devices used to change direct current into alternating current. Their power ratings range from a few milliwatts to megawatts and their frequency from a few cycles to several billion cycles per second. Oscillators are used in hundreds of applications from providing the small signal pulses for operating computer logic circuits to the heating of metal by magnetic induction.

CONSTRUCTING THE OSCILLATOR

There are several ways an oscillator can be constructed. The design is determined by several factors: the amount of power it must produce, its operating frequency, and how stable the frequency must be. Industry is making use of more and more oscillators on the production line, so the maintenance electrician should become familiar with their basic principles of operation.

APPLICATIONS OF THE OSCILLATOR

One of the most common applications of oscillators in industry is the speed control of ac induction motors. Two things determine the speed of an ac induction motor: the number of stator poles and the frequency. By changing the frequency applied to the motor, the speed of the rotating magnetic field (the *synchronous speed*) can be changed also. For instance, if a four-pole induction motor is operated at 60 hertz (Hz), the synchronous speed is 1800 revolutions per minute (rpm). If the frequency is reduced to 30 Hz, the synchronous speed will be 900 rpm. As the frequency is reduced in value, however, the inductive reactance of the motor winding is reduced also. The reduction of inductive reactance (impedance) is dealt with, however, by reducing the applied voltage as the frequency is reduced. This reduction prevents the current from becoming excessive in the stator winding and damaging the motor.

CONTROLLING THE OSCILLATOR

The circuit shown in figure 28-1 can be used to control a three-phase induction motor. The three-phase input is first changed into direct current by the three-phase bridge rectifier. The diodes of

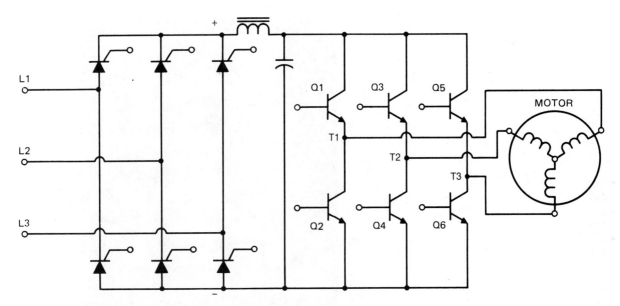

Figure 28-1 Three-phase oscillator

the bridge rectifier have been replaced by SCRs, which permit control of the output voltage. When the frequency to an induction motor is lowered, the applied voltage must be lowered. This prevents the motor from being damaged by excessive current.

Filtering

The dc output voltage is filtered before it is applied to power transistors Q1 through Q6. Power transistors Q1 through Q6 convert the dc voltage back into three-phase alternating current. This is done by firing the transistors in a specific order at specific times. Assume transistors Q1 and Q4 are switched on at the same time. This causes T1 to be positive and T2 to be negative, so current flows from T1 through the motor to T2. If transistors Q3 and Q6 are switched on at the same time, T2 is positive and T3 is negative. Current now flows from T2, through the motor, to T3. Switching the proper transistors on or off at the right time changes the dc input voltage into three-phase ac voltage. Figure 28-2 shows a variable-frequency drive.

Waveforms

The ac waveform for this oscillator is a square wave instead of a sine wave, figure 28-3. The induction motor operates without a problem on this type of waveform. Some manufacturers produce

Figure 28-2 7½-hp, 460-V ac variable frequency drive

PHOTO COURTESY OF RAMSEY CONTROLS, INC.

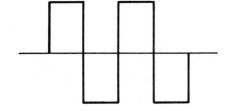

Figure 28-3 Square wave ac

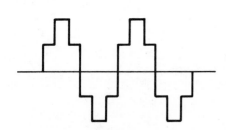

Figure 28-4 Stepped square wave

oscillators which have a stepped waveform. The stepped wave-form closer approximates the sine wave, figure 28-4.

Schematics

The oscillator shown in figure 28-1 is a basic schematic and is used to illustrate the *theory* of operation of the oscillator. The control of SCRs and transistors is generally accomplished with digital-logic circuits and a microprocessor. These will not be covered in this text.

CONVERTING FROM DC TO AC

The oscillator shown in figure 28-5 is much simpler than the one just described. It is used to convert direct current into single-

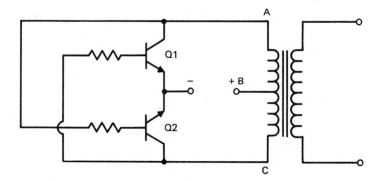

Figure 28-5 Single-phase square oscillator

phase alternating current. The output voltage is a square wave. The frequency is determined by the turns of the transformer and the applied voltage.

This circuit will operate because of the change of the magnetic structure in the core of the transformer. When the power is first applied, one transistor will begin to conduct before the other. This is due to slight imbalances in the characteristics of the transistors. It will be assumed that transistor Q1 will conduct first. When transistor Q1 turns on, current will flow from point B to point A, through Q1 and back to the negative of the power supply. As current flows through the transformer winding, a magnetic field is produced, which causes the iron molecules in the core to align themselves in one direction. When all the molecules have become aligned the transformer core becomes saturated. This causes the voltage between points B and A to drop to a very low value, causing transistor Q1 to turn off.

Transistor Q2 now turns on and permits current to flow from point B to point C and through transistor Q2 to the negative terminal of the power supply. Since current is now flowing through the winding in an opposite direction, the magnetic field reverses polarity. This causes the molecules of iron in the transformer core to realign themselves in the opposite direction. When they have all become aligned the transformer core again becomes saturated and the voltage between points B and C drops to a low value, causing transistor Q2 to turn off. Transistor Q1 again turns on and the process starts again.

Each time current flows through the primary of the transformer, it induces a voltage into the secondary. Since the current flows first in one direction and then in the other, the induced voltage is ac. The transistors in this circuit are either completely on or completely off. The output voltage is square wave alternating current.

·CONVERTING DC INTO AC USING SCRs

Due to their ability to handle large amounts of power, SCRs are often used for converting direct current into alternating current. An example of this type of circuit is shown in figure 28-6. In this circuit the SCRs are connected to a control unit that controls the sequence and rate at which the SCRs are gated on. The circuit is so constructed that SCRs A and A' are gated on at the same time and SCRs B and B' are gated on at the same time. Inductors L_1 and L_2 are used for filtering and wave shaping. Diodes D_1 through D_4 are clamping diodes and are used to prevent the output voltage from becoming excessive. Capacitor C_1 is used to turn one set of SCRs off when the other set is gated on. This capacitor must be a true ac capacitor since it will be charged to the alternate polarity each half cycle. In an inverter intended to handle large amounts of power, capacitor C_1 will be a bank of capacitors. To understand the operation of this circuit, assume that SCRs A and A' are gated on at the same time. Current will flow through the circuit as shown in figure 28-7. Notice the direction of current flow through the load and that capacitor C_1 has been charged to the polarity shown. Recall that when an SCR has been turned on by the gate, it can only be turned off by permitting the current flow through

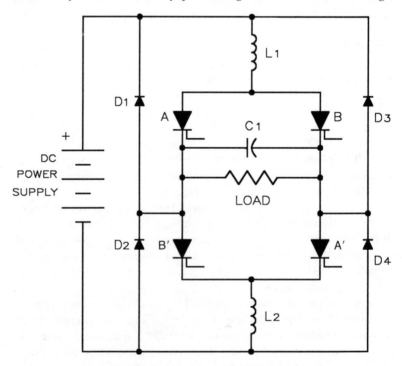

Figure 28-6 Changing dc into ac using SCRs

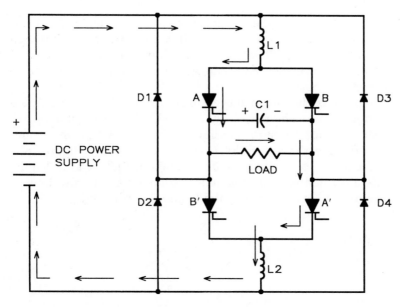

Figure 28-7 Current flows through SCRs A and A'.

the anode-cathode section to drop below the holding current level. Now assume that SCRs B and B' are gated on. Since SCRs A and A' are still turned on, there now exist two separate current paths through the circuit. The negative charge on capacitor C_1, however, causes the positive current to see a path more negative than the one through SCRs A and A'. The current now flows through SCRs B and B' to charge capacitor C_1 to the opposite polarity as shown

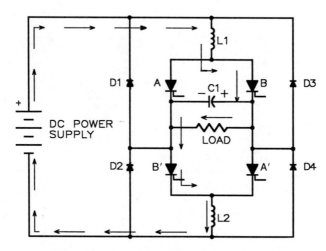

Figure 28-8 Current flows through SCRs B and B'.

in figure 28-8. Since the current now flows through SCRs B and B', SCRs A and A' turn off. Notice that the current flows through the load in the opposite direction, which produces alternating current through the load, and that capacitor C_1 has been charged to the opposite polarity.

To produce the next half cycle of ac current, SCRs A and A' are again gated on. The negatively charged side of capacitor C_1 will now cause the current to stop flowing through SCRs B and B' and begin flowing through SCRs A and A' as shown in figure 28-7. The frequency of the circuit is determined by the rate at which the SCRs are gated on. The frequency is limited by the rise time of current through inductors L_1 and L_2 and the value of C_1.

UNIT 29
The DC to DC Voltage Doubler

There may be occasions when it is necessary or desirable to have a higher dc voltage than is available. If a higher ac voltage was needed, it would be a simple matter. A transformer with the proper turns ratio and power rating could be found and connected into the circuit. Direct current, however, cannot be transformed. It must first be converted into alternating current by the use of an oscillator. It is then raised to a higher voltage, and rectified back to direct current.

In the previous unit, an oscillator was constructed like the one shown in figure 29-1. The same basic circuit can be used to construct a circuit which will double the dc voltage applied to it. In the circuit in figure 29-2, the transformer shown is a single winding. It has been center tapped, similar to a center-tapped auto

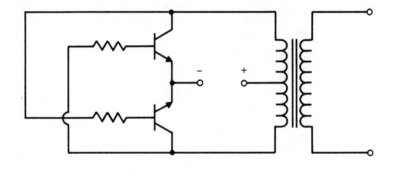

Figure 29-1 Single-phase oscillator

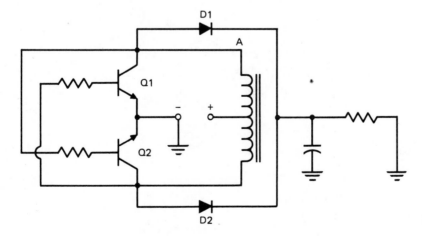

Figure 29-2 Dc voltage doubler

135

transformer. The collector of each transistor is connected to the anode of a diode. The cathodes of the diodes are connected to form a two-diode full-wave rectifier. The output of the rectifier is connected to a capacitor filter and then to the load. The load is connected to ground, the negative of the dc input.

Assume that a voltage of 6 volts dc is applied to the oscillator. When transistor Q1 turns on, current flows through half of the winding of the transformer to point A. Diode D1 is forward biased and conducts through the load to ground. Transistor Q1 turns off and Q2 turns on. Current flows through the other half of the winding to point C. Diode D2 is now forward biased. Current flows through the diode to the load resistor and to ground.

Since 6 volts has been applied across each half of the winding, the total output voltage is 12 volts. The oscillator produces a square wave ac voltage which is double the dc voltage applied to it. The rectifier then changes the ac voltage back into direct current.

UNIT 30
The Off-delay Timer

Timers have been used in industry since the beginning of individualized motor controls. Among these are dashpot, pneumatic, clock, and electronic timers. Regardless of the method used to achieve a time delay for the relay, timers have only one function. They delay changing their contact position when they are turned on or off.

On-delay timers delay when they are turned on, but change back immediately when turned off. They are often referred to as DOE on a schematic. This means delay on energize.

Off-delay timers, however, change their contacts immediately when turned on. They delay changing back, however, when turned off. Relay contacts are always shown on a schematic in their deenergized (off) position.

BASICS OF OPERATION

Refer to figure 30-1 which shows the NEMA symbol for an off-delay, normally open contact. When the timer is energized, the contact closes immediately just as any normal relay contact will. The contact remains closed as long as power is applied to the relay. When the relay is deenergized (turned off), the contact does not reopen immediately like a common relay contact. It remains closed for a predetermined amount of time before it reopens. Off-delay relays are often referred to on schematics as DODE. This means delay on deenergize.

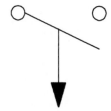

Figure 30-1 Schematic symbol of a normally open off-delay contact

Methods of Control

There are various methods used to obtain the time delay needed. This text is primarily concerned with electronic methods. There are several different ways of obtaining time delays using electronic components, but only two will be discussed. The simplest method is shown in figure 30-2.

SIMPLE TIMER: K1 is the coil of a 12-volt dc relay, and C1 is a capacitor connected in parallel with the coil. When switch S1 is closed, K1 immediately energizes. Capacitor C1 is charged to 12 volts dc. When switch S1 is opened, capacitor C1 begins to discharge through K1. This keeps K1 turned on as long as C1 can

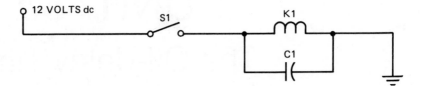

Figure 30-2 Simple capacitor timer

supply enough current to hold the relay in. This circuit uses a simple RC time constant. The amount of time K1 remains energized is determined by the resistance of coil K1 and the capacitance of C1. The circuit is useful when the time delay needed is a few seconds or less. If longer time delays are needed, the value of C1 would become very high.

ELECTRONIC TIMER: The more complex circuit shown in figure 30-3 can be used for delay times of a few seconds to an hour or

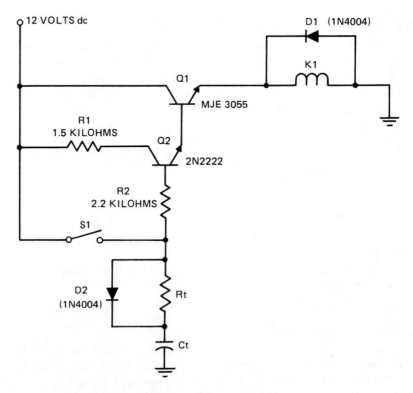

Figure 30-3 An electronic off-delay timer

more if desired. The circuit is basically a Darlington amplifier circuit. It operates as follows:

- Transistor Q1 is connected in series with the relay coil K1. It is used to turn K1 on or off.

- Transistor Q2 amplifies the base of transistor Q1.

- Resistor R1 limits the current to the base of Q1.

- Resistor R2 limits the current to the base of Q2 when switch S1 is closed.

- Switch S1 controls the operation of the circuit. When it is closed, relay K1 immediately turns on. When it is opened, relay K1 is delayed in turning off by the RC time constant of Rt and Ct.

- Diode D1 is known as a kick-back or free-wheeling diode. Its job is to kill the spike voltage created when the current through K1 is suddenly stopped. (Notice that diode D1 is connected in the circuit in the reverse bias position.)
  1. When current flows through K1, a magnetic field is developed around the coil.
  2. If the current is suddenly interrupted, the collapsing magnetic field induces a high voltage in the coil. The induced voltage can be a spike of several hundred volts. A spike voltage this large can destroy electronic components throughout the circuit.
  3. Diode D1 is reverse biased to the applied voltage of the circuit. (Note: An induced voltage is always opposite in polarity to the applied voltage.)
  4. That makes D1 forward biased to any spike voltages induced in the coil of K1.
  5. The forward voltage drop of a silicon diode is only about .7 volt.
  6. Diode D1 will not permit the induced voltage of K1 to become greater than its forward voltage drop.

- Diode D2 is used to bypass resistor Rt so capacitor C1 can be charged immediately when switch S1 is closed. If diode D2 was removed from the circuit, capacitor Ct would have to be charged through resistor Rt. This could take several minutes, depending on the values of Rt and Ct. When switch S1 is opened, the charge of capacitor Ct makes diode D2 reverse biased. Ct therefore discharges through resistor Rt.

- Capacitor Ct supplies current to the base of transistor Q2 after S1 has been opened.

- Relay K1 remains turned on as long as capacitor C1 can supply enough current to the base of Q2 to keep the circuit turned on.

- Resistor Rt determines the discharge time of Ct. The values of Ct and Rt determine the delay time of the circuit.

UNIT 31

The On-delay Timer

The on-delay timer is used throughout industry in many applications. The time delay can be accomplished by several methods: pneumatic, electric clock, or electronic. A quartz clock is used in some electronic timers which must be very accurate. Most electronic timers, however, use an RC time constant to do the job. They are inexpensive and relatively accurate. A variable resistor added to the circuit allows adjustment of the length of time delay within the limits of the circuit.

Regardless of the method used, all on-delay timers operate basically the same way. When an on-delay relay is energized, its contacts remain in their normal position for some length of time before they change. When the relay is deenergized, the contacts return to their normal position immediately.

Figure 31-1 shows the NEMA (National Electrical Manufacturers Association) symbol for a normally open set of contacts operated by an on-delay relay. Assume that this contact is controlled by an on-delay relay which has been set for 10 seconds. When the relay is energized, the contact remains open for 10 seconds and then closes. When the relay is deenergized, the contact returns to its open position immediately.

The circuit shown in figure 31-2 is a simple on-delay time circuit. It uses a unijunction transistor as the timing element. The values of Rt and C1 determine the timing of the circuit. Resistor R1 limits the current through the UJT. R2 permits a positive pulse to be produced across the resistor when the UJT turns on and discharges capacitor C1. If resistor R2 was removed and the circuit grounded, a positive pulse could not be produced when C1 is discharged.

Capacitor C2 isolates the gate of the SCR from the UJT. Any leakage current through the UJT is blocked by C2. A large pulse caused by the discharge of C1 is passed across C2 like an ac voltage would be.

Resistor R3 keeps the gate of the SCR at a ground potential until a pulse is transmitted through capacitor C2 to fire the gate of the SCR. When the SCR fires, it provides current to K1; the coil of a 12-volt dc relay coil. Once the SCR fires, it remains turned on until the circuit is broken by switch S1. Diode D1 is used as a kickback or free-wheeling diode. It kills the voltage spike induced in the coil of K1 when the circuit is opened.

OBJECTIVES

After studying this unit the student should be able to:

- Discuss the operation of an on-delay timer
- Draw the schematic symbol for an on-delay timer contact
- Construct an on-delay timer using discrete electronic components

Figure 31-1 Schematic symbol of a normally open on-delay contact

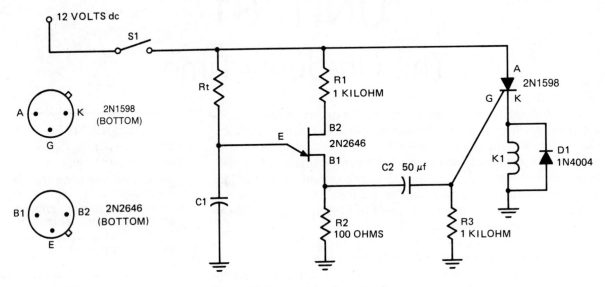

Figure 31-2 An on-delay timer circuit

UNIT 32
The Pulse Timer

The pulse timer is similar in construction to the on-delay timer, with one major difference. When the on-delay timer is turned on, it remains that way until the power is interrupted. The pulse timer, however, turns itself back off after it has been turned on. The windshield wiper delay circuit on many automobiles is a good example of this type of circuit.

The circuit shown in figure 32-1 is a simple pulse-timer circuit. The unijunction transistor, Q1, is used to provide a pulse. The time between pulses is determined by the RC time constant of R1 + R2 and C1. When the voltage across capacitor C1 becomes high enough, the UJT turns on and discharges C1 through resistor R6 to ground. As the capacitor discharges through R6, a positive voltage spike appears across R6.

Transistor Q3 is connected in series with K1 which is the coil of a twelve-volt dc relay. Resistor R5 limits the base current for transistor Q3. When transistor Q3 turns on, coil K1 turns on also. Capacitor C3 is connected to the base of transistor Q3. It is used to keep the base of transistor Q3 at ground until a positive voltage appears at the base of Q3.

OBJECTIVES

After studying this unit the student should be able to:
- Discuss the operation of a pulse timer
- Construct a pulse timer using discrete electronic components

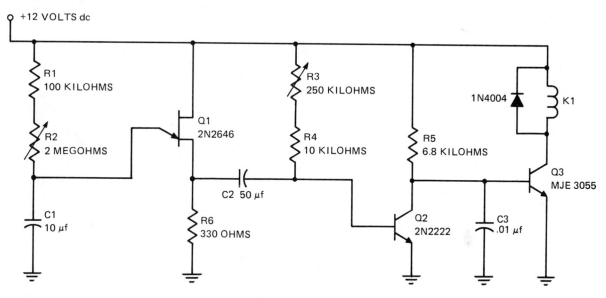

Figure 32-1 Electronic pulse timer

Transistor Q2 is used as a stealer transistor. A stealer transistor *steals* the base current from another transistor to keep it turned off. In this circuit, Q2 steals the current from the base of transistor Q3 to keep it turned off. As long as transistor Q2 is turned on this condition will continue. The resistance of R3 + R4 determines the amount of base current to transistor Q2. Resistor R4 ensures that the current to the base of Q2 will be limited if the value of resistor R3 is adjusted to 0 ohm.

Capacitor C2 is charged through the resistance of R3 + R4. When capacitor C2 is charged sufficiently, base current keeps transistor Q2 turned on. When transistor Q2 is turned on, the base of transistor Q3 is at ground potential and therefore off.

When the UJT turns on and discharges capacitor C1, a large positive pulse is produced at capacitor C2. This large positive pulse causes the base-emitter junction of transistor Q2 to become forward biased. Electrons flow from ground through the base-emitter junction to the capacitor. Since electrons are negative particles, the base of transistor Q2 becomes negative which turns the transistor off. This permits base current to flow to transistor Q3 and turn on relay K1. Transistor Q2 will remain turned off until the capacitor C2 can again be charged positive through resistors R3 and R4. In actual circuit operation, resistor R2 determines the time between pulses, and resistor R3 determines the amount of time the relay stays turned on.

UNIT 33
The 555 Timer

The 555 timer has become one of the most popular electronic devices used in industrial electronics circuits. The reason for the popularity is the 555's tremendous versatility. It is used in circuits requiring a time-delay function and also as an oscillator to provide the pulses needed to operate computer circuits.

PIN EXPLANATION

The 555 timer is an eight pin inline integrated circuit (IC), figure 33-1. This package will have a notch at one end or a dot near one pin. These identify pin #1. Once pin #1 has been identified, the other pins are numbered as shown. The following is an explanation of each pin and what it does.

Pin #1 **Ground**—This pin is connected to circuit ground.

Pin #2 **Trigger**—This pin must be connected to a voltage which is less than one-third Vcc (the applied voltage) to trigger the unit. This is generally done by connecting pin #2 to ground. The connection to one-third Vcc or ground must be momentary. If pin #2 is not removed from ground the unit will not operate.

Pin #3 **Output**—The output turns on when pin #2 is triggered, and off when the discharge is turned on.

Pin #4 **Reset**—When this pin is connected to Vcc it permits the unit to operate. When connected to ground it activates the discharge and keeps the timer from operating.

Pin #5 **Control voltage**—If this pin is connected to Vcc through a variable resistor, the on time becomes longer. The off time remains unaffected. If it is connected to ground through a variable resistor, the on time becomes shorter. The off time is still not affected. If it is not used in the circuit, it is generally taken to ground through a small capacitor. This prevents circuit noise from *talking* to pin #5.

Pin #6 **Threshold**—When the voltage across the capacitor connected to pin #6 reaches two-thirds the value of Vcc, the discharge turns on and the output turns off.

Pin #7 **Discharge**—When pin #6 turns the discharge on, it discharges the capacitor connected to pin #6. The discharge

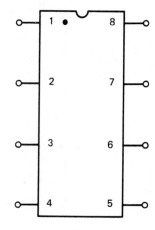

Figure 33-1 Eight pin integrated circuit

remains turned on until pin #2 retriggers the timer. It then turns off and the capacitor connected to pin #6 begins charging again.

Pin #8 **Vcc**—This pin is connected to the applied voltage of the circuit which is known as Vcc. The 555 timer operates on a wide range of voltages. Its operating voltage range is considered to be between 3 and 16 volts dc.

(Note: For the following explanation, assume that pin #2 is connected to pin #6. This permits the unit to be retriggered by the discharge each time it turns on and discharges the capacitor to one-third the value of Vcc.)

OPERATION OF THE TIMER

The 555 timer operates on a percentage of the applied voltage. The time setting remains constant even if the applied voltage changes. When the capacitor connected to pin #6 reaches two-thirds of the applied voltage, the discharge turns on. The capacitor discharges until it reaches one-third of the applied voltage. If Vcc of the timer is connected to 12 volts dc, two-thirds of the applied voltage is 8 volts and one-third is 4 volts. This means that when the voltage across the capacitor connected to pin #6 reaches 8 volts, pin #7 turns on. After the capacitor discharges to one-third the value of Vcc, or 4 volts, it turns off, figure 33-2.

If the voltage is lowered to 6 volts at Vcc, two-thirds of the applied voltage becomes 4 volts and one-third is 2 volts. Pin #7 now turns on when the voltage across the capacitor connected to pin #6 reaches 4 volts. It turns off when the voltage across the capacitor drops to 2 volts. The formula for an RC time constant is: time = resistance × capacitance. There is no mention of voltage in the formula. It takes just as long to charge the capacitor if the circuit is connected to 12 volts as it will for 6 volts. The voltage of the capacitor connected to pin #6 reaches two-thirds of Vcc the same way. When the timer has an applied voltage of 12 volts it takes the same amount of time as when the applied voltage is only 6 volts. Notice that the timing of the circuit remains the same even if the voltage changes.

Using a Circuit to Explain the Timer

The circuit shown in figure 33-3 is used to help explain the operation of the 555 timer. The figure shows a normally closed switch, S1, connected between the discharge, pin #7, and the ground pin #1. There is a normally open switch, S2, connected between the output, pin #3, and Vcc at pin #8. The dotted line

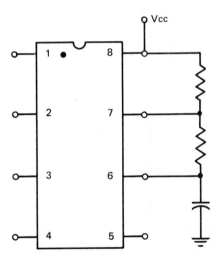

Figure 33-2 RC time constant controls the operation of the timer.

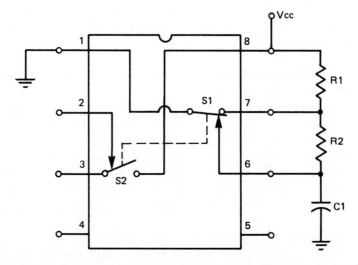

Figure 33-3 Basic timer operation

drawn between these two switches shows mechanical connection: both switches operate together. If S1 opens, S2 closes at the same time. If S2 opens, S1 closes. Pin #2, the trigger, and pin #6, the threshold, are used to control these two switches. The trigger can close switch S2, and the threshold can close S1. To begin this analysis, assume switch S1 is closed and switch S2 is open as shown in figure 33-3.

1. When the trigger is connected to a voltage less than one-third of Vcc, switch S2 closes and switch S1 opens.
2. When switch S2 closes, voltage is supplied to the output at pin #3.
3. When switch S1 opens, the discharge is no longer connected to ground.
4. Capacitor C1 begins to charge through resistor R1 and R2.
5. When the voltage across C1 reaches two-thirds of Vcc, the threshold, pin #6, causes switch S1 to close and switch S2 to open.
6. When switch S2 opens, the output turns off.
7. When switch S1 closes, the discharge, pin #7, is connected to ground.
8. This permits capacitor C1 to begin discharging through resistor R2.
9. The state of the timer will remain in this position until the trigger is again connected to a voltage which is less than one-third of Vcc.

If the trigger is permanently connected to a voltage less than one-third of Vcc, switch S2 is held closed and switch S1 is held open. This, of course, stops the operation of the timer. The trigger must be a momentary pulse and not a continuous connection if the 555 is to operate.

APPLICATIONS

Thousands of uses for the 555 timer have been found in both commercial and industrial electronics. One application is shown in figure 33-4. In this circuit, the 555 timer is the heart of a continuity tester that gives both a visual and an audible indication of when a complete circuit exists. When the test probes are connected together, visual proof of a complete circuit is provided by the light-emitting diode, D_1. Resistor R_1 limits current flow through the LED.

The audible part of the tester is provided by the 555 timer and the small speaker. When the test probes are connected together, power is provided to the timer. Since pin #2 has been connected directly to pin #6, the timer will operate in an astable mode and provide a pulsating output at pin #3. Capacitor C_2 changes the pulsating dc into ac to operate the speaker. Resistor R_2 limits the current flow when the discharge pin, #7, turns on. The tone of the speaker can be adjusted by changing the value of resistor R_3 or capacitor C_1.

Another example of how a 555 timer can be used is shown in

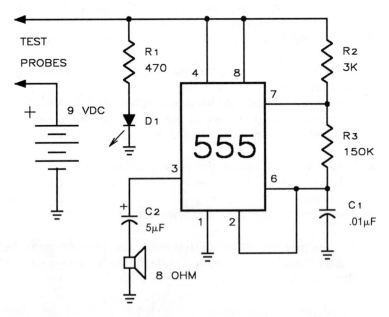

Figure 33-4 Continuity tester

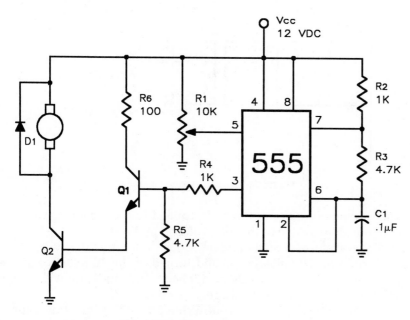

Figure 33-5 DC motor speed control

figure 33-5. In this circuit, the timer is used to provide speed control to a direct current motor. The controller works by changing the amount of voltage applied to the armature of the motor. The 555 timer provides a pulsating dc voltage to the armature of the motor. The voltage applied across the armature is the average value determined by the length of time transistor Q_2 is turned on as compared to the length of time it is turned off. For example, the waveform in figure 33-6 has a peak value of 12 volts. Notice the voltage is off three times longer than it is on. Assume the interval between pulsations to be 40 milliseconds. If the voltage applied to the armature is turned on for 10 milliseconds and turned off for 30

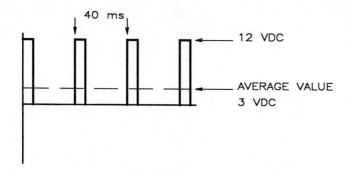

Figure 33-6 Average dc voltage is 3 volts.

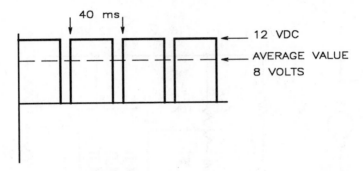

Figure 33-7 Average voltage is 8 volts.

milliseconds, the average voltage applied to the armature will be one-fourth the value of the total voltage, or 3 volts.

The waveform shown in figure 33-7 indicates that the voltage is turned on for 30 milliseconds and turned off for 10 milliseconds. The average amount of voltage applied to the armature will now be three-fourths the total voltage, or 8 volts.

The 555 timer is used to control the amount of time transistor Q_2 is turned on or off. The timer is operated in the astable mode by connecting pin #2 to pin #6. Resistor R_2 limits the amount of current flow when the discharge turns on. Resistor R_3 and capacitor C_1 determine the amount of off time for the timer. Resistor R_4 limits current flow to the base of transistor Q_1, which is used as a Darlington driver for transistor Q_2. Resistor R_5 ensures transistor Q_1 remains turned off when there is no output at pin #3 of the timer. Resistor R_6 limits base current to transistor Q_2 when transistor Q_1 turns on. Diode D_1 performs the function of a commutating diode to prevent voltage spikes being produced when current flow through the armature stops. Resistor R_1 controls the length of time the output of the timer will be turned on, which controls the speed of the motor. If the wiper of resistor R_1 is adjusted close to Vcc, the output will be turned on for a long period of time as compared with the amount of time it will be turned off. If the wiper is adjusted close to ground, the output will be turned on for only a short period of time as compared with the off period.

UNIT 34

The 555 Timer Used as an Oscillator

The 555 timer can perform a variety of functions. One of the most common functions is that of an oscillator. The 555 timer has become popular for this application because it is so easy to use, figure 34-1.

OPERATION OF THE TIMER

The 555 timer shown in figure 34-1 has pin #2 connected to pin #6. This permits the timer to retrigger itself at the end of each time cycle. When the voltage at Vcc is first turned on, capacitor C1 discharges and has a voltage of 0 volt across it. Since pin #2 is connected to pin #6, and the voltage at that point is less than one-third Vcc, the timer triggers. When the timer triggers, two things happen at the same time: the output turns on and the discharge turns off. When the discharge at pin #7 turns off, capacitor C1 begins charging through resistors R1 and R2. The time it takes for capacitor C1 to charge is determined by its capacitance and the combined resistance of R1 + R2. When capacitor C1 is charged to a voltage which is two-thirds of Vcc, two things happen: the output turns off, and the discharge at pin #7 turns on. When the

OBJECTIVES

After studying this unit the student should be able to:

- Discuss the operation of a 555 timer when it is used as an oscillator
- Connect a 555 timer as an oscillator
- Make measurements of frequency using an oscilloscope

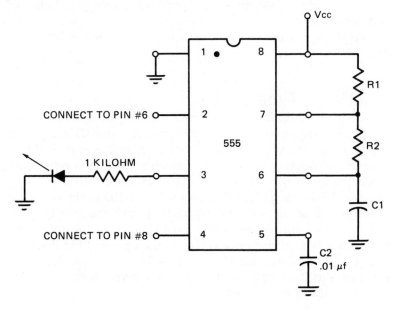

Figure 34-1 A stable oscillator

151

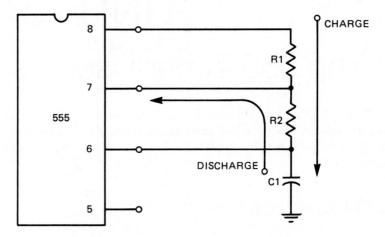

Figure 34-2 The charge time is longer than the discharge time.

discharge turns on, capacitor C1 discharges through resistor R2 to ground. The time it takes C1 to discharge is determined by the capacitance of C1 and the resistance of R2. When capacitor C1 has discharged to where its voltage is one-third of Vcc, the timer is retriggered by pin #2. When the timer is retriggered, the output again turns on and the discharge turns off. When the discharge turns off, capacitor C1 again begins to charge.

Notice that the amount of time required to charge capacitor C1 is determined by the combined resistance of R1 + R2. The discharge time, however, is determined by the value of R2, figure 34-2.

Discharge Time

The timer's output is turned on during the period of time capacitor C1 is charging. It is turned off during the time C1 is discharging. The on time of the output, therefore, is longer than the off time. If the value of resistor R2 is much greater than resistor R1, this condition is not too evident. For instance, if resistor R1 has a value of 1 kilohm (k) and R2 has a value of 100 k, the resistance connected in series with the capacitor during charging is 101 k. The resistance connected in series with the capacitor during discharge is 100 k. In this circuit, the charge time and the discharge time of the capacitor are 1% of each other. If an oscilloscope is connected to the output of the timer, a waveform similar to that shown in figure 34-3 is seen.

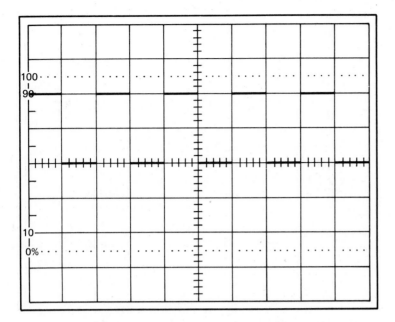

Figure 34-3 On time and off time are the same.

Assume that the value of resistor R1 is changed to 100 k and that the value of resistor R2 remains at 100 k. In this circuit, the resistance connected in series with the capacitor during charging is 200 k. The resistance connected in series with the capacitor during discharge, however, is 100 k. In the circuit, the discharge time is 50% of the charge time. This means that the output of the timer is turned on twice as long as it is turned off. An oscilloscope connected to the output of the timer displays a waveform similar to the one shown in figure 34-4.

USING PIN #5: Although this condition can exist, the 555 timer has a provision for solving the problem. Pin #5, the *control voltage* pin, can give complete control of the on time of the timer. Pin #5 will not affect the off time, but it does give control of the on time. The off time is controlled by the amount of capacitance of C1 and the resistance of R2. If pin #5 is connected to Vcc through a resistor, the amount of on time will be greater and the off time will remain the same. If pin #5 is connected to ground through a resistor, the on time will be shorter and the off time will remain the same. Complete control of the on time can be gained by using a variable resistor as a potentiometer and connecting pin #5 to the

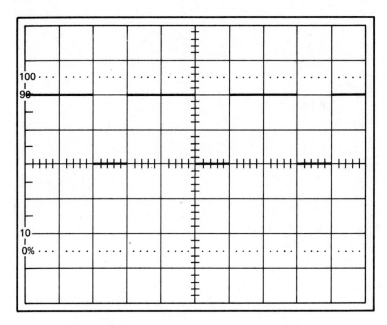

Figure 34-4 On time is longer than off time.

wiper of the pot as shown in figure 34-5. In this circuit, the voltage applied to pin #5 can be varied between Vcc and ground depending on the position of the wiper. This will give complete control of the on time of the timer.

SUMMARY

- The output frequency of the timer is determined by the values of capacitor C1 and resistors R1 and R2.
- It will operate at almost any frequency desired.
- It has found use in many industrial electronic circuits which require the use of a square wave oscillator.

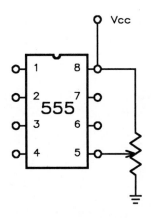

Figure 34-5 Potentiometer controls length of pulse.

UNIT 35
The 555 On-delay Timer

The 555 timer can be used to construct an on-delay relay. The 555 provides accurate time delays which can range from seconds to hours. These delays depend on the values of resistance and capacitance used in the circuit.

In the circuit shown in figure 35-1, transistor Q1 is used to switch relay coil K1 on or off. The 555 timer may not be able to supply the current needed to operate the relay, so a transistor will be used to do the job of controlling the relay.

OPERATION OF THE TIMER

Transistor Q2 is used as a stealer transistor to steal the base current from transistor Q1. As long as transistor Q2 is turned on by the output of the timer, transistor Q1 is turned off.

OBJECTIVES

After studying this unit the student should be able to:
- Describe the operating characteristics of an on-delay timer
- Discuss the operation of a stealer transistor
- Construct a circuit using the 555 timer as an on-delay timer

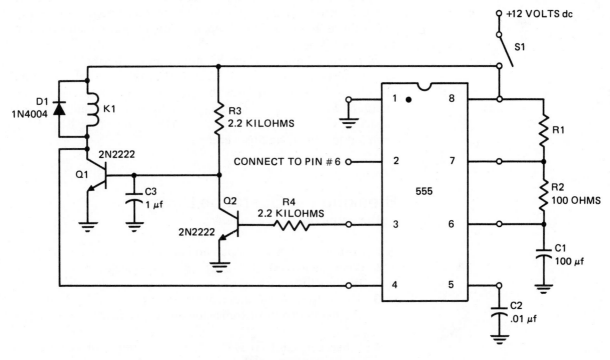

Figure 35-1 On-delay timer

Capacitor C3 is connected from the base of transistor Q1 to ground. It acts as a short time-delay circuit. When Vcc is first turned on by switch S1, capacitor C3 is in a discharged state. Before transistor Q1 can turn on, capacitor C3 must be charged through resistor R3. This charging time is only a fraction of a second. It ensures that transistor Q1 will not turn on before the output of the timer can turn on transistor Q2. Once transistor Q2 is turned on, it holds transistor Q1 off by stealing its base current.

Diode D1 is used as a kick-back or free-wheeling diode. It kills the spike voltage induced into the coil of relay K1 when switch S1 is opened. Resistor R3 limits the base current to transistor Q1. Resistor R4 limits the base current to transistor Q2.

Pin #4 is used as a latch in this circuit. When power is first applied at Vcc, transistor Q1 is turned off. Since Q1 is off, most of the applied voltage will be dropped across the transistor. This makes about 12 volts appear at the collector of the transistor. Pin #4 is connected to the collector so 12 volts is applied to pin #4. Pin #4 (the *reset* pin) must be connected to a voltage greater than two-thirds of Vcc if the timer is to operate. When it is connected to a voltage less than one-third of Vcc, it turns on the discharge and keeps the timer from operating. When transistor Q1 turns on, the collector of the transistor drops to ground or 0 volt. Pin #4 is also connected to ground, preventing the timer from further operation. Since the timer cannot operate, the output remains off, and transistor Q1 remains on.

Capacitor C1 and resistors R1 and R2 are used to set the amount of time delay. (Note: Resistor R2 should be kept at a value of about 100 ohms. Its job is to limit the current when capacitor C1 discharges.) Resistor R2 was made a relatively low value to enable capacitor C1 to discharge quickly. The value of resistor R1 is changed to adjust the time setting.

Breaking the Operation Down

The circuit operates as follows:

1. Assume switch S1 is open and all capacitors are discharged.
2. When switch S1 closes, pin #2 is connected to 0 volt and triggers the timer.
3. The output of the timer turns on transistor Q2 which steals the base current of transistor Q1. Q1 remains off as long as Q2 is on.
4. When capacitor C1 is charged to two-thirds of Vcc, the discharge turns on and the output of the timer turns off.

5. When the output turns transistor Q2 off, transistor Q1 is supplied with base current through resistor R3 and turns on relay coil K1.
6. With transistor Q1 on, the voltage applied to the reset pin, #4, changes from 12 volts to 0 volt.
7. When the reset is taken to 0 volt, the discharge is locked on and the output off.
8. Once transistor Q1 has turned on, switch S1 has to be reopened to reset the circuit.

UNIT 36
The 555 Pulse Timer

OBJECTIVES

After studying this unit the student should be able to:

- Describe the operation of a pulse timer
- Discuss the operation of a blocking diode
- Construct a pulse timer using a 555 timer

In the circuit of figure 36-1, the 555 is used as a pulse timer. It is connected differently than the other 555 timer circuits which have been covered. In this circuit, capacitor C1 is charged by a feed-back circuit from the output of the timer and not from Vcc.

- Transistor Q1 is used to turn relay coil K1 on or off.
- Diode D1 is used as a kick-back or free-wheeling diode.
- Resistor R1 limits the current to the base of transistor Q1.
- Capacitor C2 is used to disable pin #5 to ground.
- Diode D2 is a blocking diode.
- Once capacitor C1 has been charged, it is prevented from discharging through resistors R2 and R1; it must discharge through resistors R3 and R4.

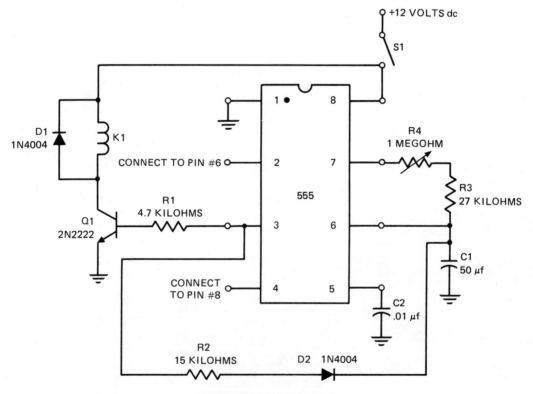

Figure 36-1 Pulse timer

In this circuit pin #2 is connected to pin #6 which is at ground potential. This voltage is less than one-third of Vcc, so the output at pin #3 turns on when switch S1 closes. With pin #3 on, two things happen: transistor Q1 turns on and energizes relay coil K1 and capacitor C1 begins charging through resistor R2. When capacitor C1 is charged to two-thirds of Vcc, the output turns off, which turns off relay K1. The value of resistor R2 determines how long relay K1 stays on before it turns off.

When capacitor C1 has been charged to two-thirds of Vcc, and the output has turned off, it discharges through resistors R3 and R4. When capacitor C1 has been discharged to a point that its voltage is one-third of Vcc, pin #2 again triggers and turns the output on. The value of resistors R3 + R4 determines how long relay coil K1 remains off before it turns on again. If the time between pulses is a fixed value, resistor R4 is not necessary. If R4 is a variable resistor, however, it will control the time between pulses.

UNIT 37

Above- and Below-ground Power Supplies

OBJECTIVES

After studying this unit the student should be able to:

- Describe an above- and below-ground power supply
- Discuss different methods of constructing an above- and below-ground power supply
- Discuss filtering for an above- and below-ground power supply
- Construct an above- and below-ground power supply

Most electronic circuits require a dc voltage to operate. A power supply which can furnish a positive and negative output voltage is sufficient for most circuits. However, there are some which require voltages that are above- and below-ground.

DEFINING THE TERMS

In an electronic circuit, ground is considered to be 0 volt. Other values of voltage are *referenced* to ground. If a voltage is positive with respect to ground, it is considered to be *above* ground. If a voltage is negative with respect to ground, it is considered to be *below* ground, figure 37-1. Voltmeters V1 and V2 are both zero center voltmeters. Both have their negative terminals connected to ground and their positive terminals connected to an input line.

Applying the Definitions

Notice that the pointer on voltmeter V1 has moved to the right of the zero center. This indicates that the + (positive) terminal of the voltmeter is connected to a voltage which is positive with respect to ground. The pointer of voltmeter V2, however, has moved to the left of the zero center, which indicates that its + terminal is connected to a voltage which is negative with respect

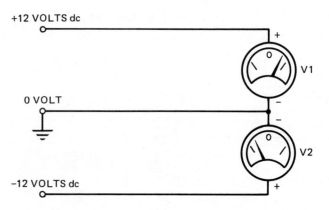

Figure 37-1 Above- and below-ground voltages

160

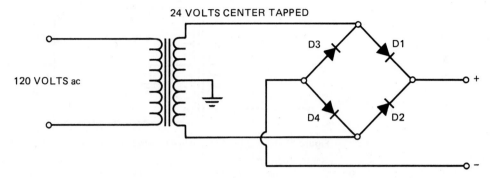

Figure 37-2 An above- and below-ground power supply

to ground. If the + line is 12 volts above ground and the −
(negative) line is 12 volts below ground, a voltmeter connected
from the + line to the − line will indicate 24 volts.

CONSTRUCTION OF THE SUPPLY

Center-tapped

There are several ways to construct an above- and below-
ground power supply. One of the easiest is shown in figure 37-2.
The circuit is constructed by using a center-tapped transformer
and a bridge rectifier. Examine this circuit and notice that there
are 2 two-diode type full-wave rectifiers. Diodes D1 and D2 form
one rectifier which produces a positive voltage with respect to
ground, figure 37-3. Diodes D3 and D4 form a two-diode type full-
wave rectifier which produces a negative voltage when compared
to ground, figure 37-4.

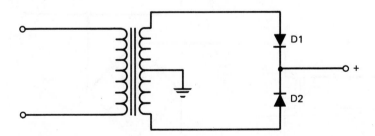

Figure 37-3 A two diode full-wave rectifier producing a + (plus) voltage when
compared to ground.

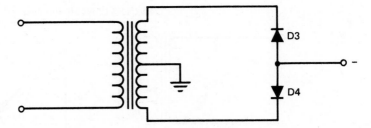

Figure 37-4 A two diode full-wave rectifier producing a − (minus) voltage when compared to ground.

The above- and below-ground power supply shown in figure 37-2 produces voltages which are equal when compared to ground. For instance, if the positive voltage is 12 volts above ground, the negative voltage is 12 volts below ground. Another method for producing an above- and below-ground power supply with equal voltages is shown in figure 37-5.

Two Transformers

This circuit uses two separate transformers and two bridge rectifiers. The negative output voltage of rectifier #1 is connected to the positive output voltage of rectifier #2. The point where this connection is made is the ground terminal. The positive output of rectifier #1 is the positive or above-ground voltage. The negative

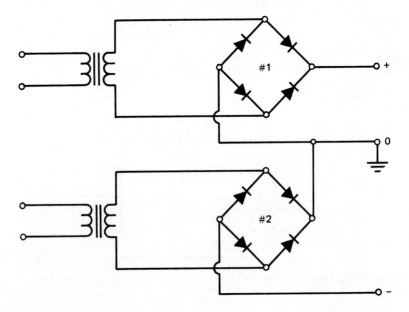

Figure 37-5 A two-bridge rectifier circuit

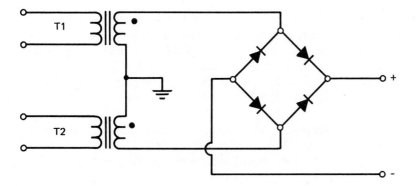

Figure 37-6 Two transformers connected to form a center-tapped winding

output of rectifier #2 is the negative or below-ground voltage. This circuit can be used when a center-tapped transformer is not available, or when a higher voltage is needed than is available with one transformer.

The circuit in figure 37-5 illustrates a method of obtaining an above- and below-ground power supply using two transformers. The same results can be achieved by a simpler method, however.

Series Aided

Instead of using two bridge rectifiers, the secondary winding of the transformers can be connected series aiding as shown in figure 37-6. The connection point of the two transformer secondaries can be used as the center tap. Only one bridge rectifier is required in this case. This circuit operates the same as the circuit shown in figure 37-2.

For small power applications, the circuit shown in figure 37-7 can be used to obtain above- and below-ground voltages. Diodes D1 and D2 are zener diodes of equal value. The applied dc voltage

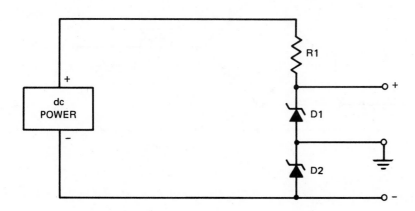

Figure 37-7 Above- and below-ground voltage produced by zener diode

must be greater than the combined voltage of the two zener diodes. For instance, if the zeners are rated at 12 volts each, the applied voltage must be greater than 24 volts. Resistor R1 limits the current flow through the circuit. Its value must be set to permit enough current flow to operate the load. Since each zener diode has a voltage drop of 12 volts,

1. the connection point of the two diodes will be 12 volts more negative than the cathode of diode D1,
2. and 12 volts more positive than the anode of diode D2.

If the connection point of the two diodes is grounded, the cathode of diode D1 is positive or above ground, and the anode of diode D2 is negative or below ground.

FILTERING

Above- and below-ground power supplies are filtered the same way as any other power supply. A choke coil can be used to filter the current, and a capacitor to filter the voltage, figure 37-8.

Notice the connection of capacitor C2 in figure 37-8. The positive terminal has been connected to ground. Since the ground terminal of the power supply is more positive than the negative terminal, the polarity of the capacitor must be observed when this connection is made.

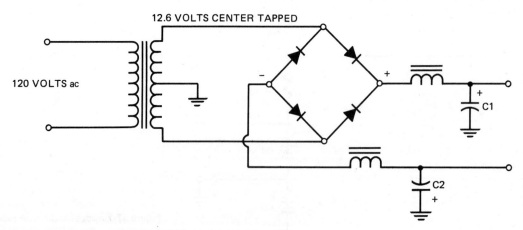

Figure 37-8 Filtering an above- and below-ground power supply

UNIT 38
The Operational Amplifier

The operational amplifier, like the 555 timer, has become another very common component found in industrial electronic circuits. The operational amplifier *(op amp)* is used in hundreds of different applications. Op amps differ, depending on the circuit which it is intended to operate. Some use bipolar transistors and others use field effect transistors (FETs) for the input. Field effect transistors have extremely high input impedance. This can be several thousand megohms. The advantage of this impedance is that a large amount of current is not needed to operate the amplifier. Op amps which use FET inputs are generally considered as requiring no input current.

OBJECTIVES
After studying this unit the student should be able to:
- Discuss the operation of an operational amplifier
- Discuss inverting inputs and noninverting inputs
- Describe specific parameters for the 741 operational amplifier
- Discuss the operation of the offset null
- Discuss negative feedback and calculate the gain of the operational amplifier
- Connect an operational amplifier in a circuit

THE IDEAL OP AMP

There are three things that would go into the making of an *ideal* amplifier. These are:

1. The ideal amplifier should have an input impedance of infinity. If the amplifier has such an impedance, it requires no power drain on the signal source that is to be amplified. Therefore, regardless of how weak the input signal source is, it will not be affected when connected to the amplifier.
2. The ideal amplifier should have zero output impedance. If it has zero output impedance, it can be connected to any load resistance desired. No voltage will drop inside the amplifier. If there is no internal voltage drop, the amplifier will utilize 100% of its gain.
3. The amplifier will have unlimited gain. This permits it to amplify any input signal as much as desired.

THE 741 OP AMP

There is no such thing as the ideal or perfect amplifier, of course, but the op amp comes close. One of the op amps which is still used in industry is the 741. It will be used in this unit as a typical operational amplifier. There are other op amps which have different characteristics of input and output impedance, but the basic theory of operation is the same for all.

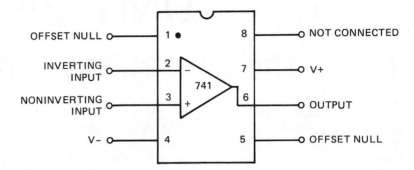

Figure 38-1 The 741 operational amplifier

The 741 op amp uses bipolar transistors for the input. The input impedance is about 2 megohms, and the output impedance is about 75 ohms. Its open loop (maximum gain) is about 200,000. The 741 op amp has such a high gain that it is generally impractical to use. Negative feedback, which will be discussed later in the unit, is used to reduce the gain. Assume the amplifier has an output voltage of 15 volts. If the input signal voltage is greater than 1/200,000 of the output voltage or 75 microvolts, (15/200,000 = .000075), the amplifier will be driven into saturation.

Operating the 741

The 741 operational amplifier is generally housed in an 8-pin inline IC package, figure 38-1. Pins #1 and #5 are connected to the offset null which can be used to produce 0 volt at the output. This is the process:

1. The op amp has two inputs: the inverting and the noninverting input.
2. They are connected to a differential amplifier which amplifies the difference between the two voltages.
3. If both of these inputs are connected to the same voltage (by grounding both inputs), the output should be 0 volt.

In actual practice, however, there are generally unbalanced conditions in the op amp which cause a voltage to be produced at the output. Since the op amp has a very high gain, any slight imbalance of a few microvolts at the input can cause several millivolts at the output. The offset nulls are adjusted after the 741 is connected into a working circuit. Adjustment is made by connecting a 10 kilohm (k) potentiometer across pins #1 and #5, and connecting the wiper to the negative voltage, figure 38-2.

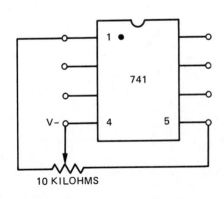

10 KILOHMS

Figure 38-2 Offset null connection

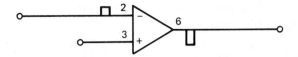

Figure 38-3 Inverted output

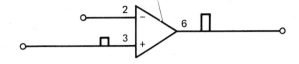

Figure 38-4 Noninverted output

Pin Explanation

Pin #2 is the inverting input. If a signal is applied to this input, the output will be inverted. A positive going ac voltage applied to the inverting input will produce a negative going output voltage, figure 38-3.

Pin #3 is the noninverting input. When a signal voltage is applied to the noninverting input, the output voltage will be the same polarity. A positive going ac signal applied to the noninverting input will produce a positive output voltage, figure 38-4.

Pins #4 and #7 are the voltage input pins. Operational amplifiers are generally connected to above- and below-ground power supplies. There are some circuit connections that do not require an above- and below-ground power supply, but these are the exception instead of the rule. Pin #4 is connected to the negative (below-ground) voltage and pin #7 is connected to the positive (above-ground) voltage. The 741 will operate on voltages that range from about 4 to 16 volts. The operating voltage is usually 12 to 15 volts plus and minus. The 741 has a maximum power output rating of about 500 milliwatts. Pin #6 is the output and pin #8 is not connected.

Open Loop Gain

As stated before, the open loop gain of the 741 operational amplifier is about 200,000. This is not practical for most applications, so this gain must be reduced to a reasonable level. One of the op amp's greatest advantages is the ease with which the gain can be controlled, figure 38-5. The amount of gain is controlled by

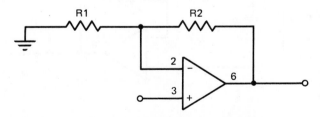

Figure 38-5 Negative feedback connection

a negative feedback loop; a portion of the output voltage is fed back to the inverting input. Since:

1. the output voltage is always opposite in polarity to the inverting input voltage,
2. the amount of output voltage fed back to the input tends to reduce the input voltage.

Negative feedback has two effects on the operation of the amplifier. It reduces the gain, and it makes the amplifier more stable.

The gain of the amplifier is controlled by the ratio of resistors R2 and R1. If a noninverting amplifier is used, the gain is found by the formula $\dfrac{(R2 + R1)}{R1}$. If resistor R1 is 1 kilohm and resistor R2 is 10 kilohms, the gain of the amplifier is 11 (11,000/1,000 = 11).

If the op amp is connected as an inverting amplifier, however, the input signal will be out-of-phase with the feedback voltage of the output. This causes a reduction of the input voltage applied and a reduction in gain. The formula R2/R1 is used to compute the gain of an inverting amplifier. If resistor R1 is 1 kilohm and resistor R2 is 10 kilohms, the gain of the inverting amplifier is 10 (10,000/ 1,000 = 10).

There are some practical limits, however. As a rule, the 741 operational amplifier is not operated above a gain of 100. If more gain is desired, it is generally obtained by using more than one amplifier, figure 38-6.

As shown in the figure, the output of one amplifier is fed into the input of another. The 741 is not operated at high gain because at high gains it tends to become unstable. Another rule for operating the 741 op amp is that the total feedback resistance, R1 + R2, is usually kept at more than 1000 and less than 100,000 ohms. These rules apply to the 741 operational amplifier but may not apply to other op amps.

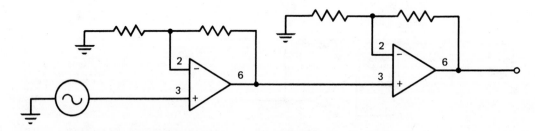

Figure 38-6 Using more than one op-amp to obtain higher gain

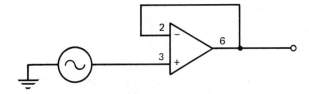

Figure 38-7 Voltage follower connection

USING THE OP AMP

Op amps are generally used in three basic ways. This does not mean that they are used in only three circuits, but that there are three basic circuits used to build other circuits.

One of these is the voltage follower. In this circuit, the output of the op amp is connected directly back to the inverting input, figure 38-7. Since there is a direct connection between the output of the amplifier and the inverting input, the gain of this circuit is one. For instance, if a signal voltage of .5 volt is connected to the noninverting input, the output voltage will be .5 volt also. This circuit amplifies the input impedance by the amount of the open loop gain. If the 741 has:

- an open loop gain of 200,000 and
- an input impedance of 2 megohms,
- the circuit will give the amplifier an input impedance of 200 k × 2 meg or 400,000 megohms.

The circuit connection is generally used for impedance matching purposes.

The second basic circuit is the noninverting amplifier, figure 38-8. In this circuit, the output voltage is the same polarity as the input voltage. If the input voltage is a positive going voltage, the output is a positive going voltage. The amount of gain is set by the ratio of resistors R1 + R2/R1 in the negative feedback loop.

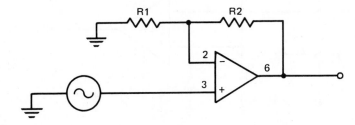

Figure 38-8 Noninverting amplifier connection

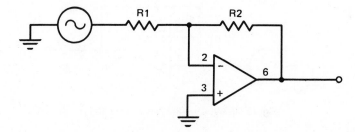

Figure 38-9 Inverting amplifier connection

The third basic circuit is the inverting amplifier, figure 38-9. In this circuit the output voltage is opposite in polarity to the input voltage. If the input signal is a positive going voltage, the output voltage is negative going at the same instant in time. The gain of the circuit is determined by the ratio of resistors R2/R1.

APPLICATIONS

In the circuit shown in figure 38-10, an operational amplifier is used as a timer. Resistors R_1 and R_2 form a voltage divider that supplies one-half the input voltage to the inverting input. Resistor R_3 and capacitor C_1 form an RC time constant. When switch S_1 is closed, the noninverting input is held at ground potential, which forces the output to remain low. When switch S_1 is opened, capacitor C_1 begins to charge through resistor R_3. When the voltage applied to the noninverting input reaches a value greater than that applied to the inverting input, the output goes high. The amount of time delay is determined by the values of C_1 and R_3. By choosing the correct values, the timer can have a delay that can range from seconds to hours. The time delay is approximately 0.7CR seconds where the value of C is in farads.

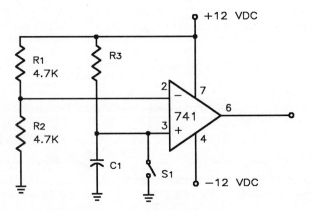

Figure 38-10 Operational amplifier used as a timer

UNIT 39
The 741 Op Amp Level Detector

The operational amplifier is often used as a level detector or comparator. In the circuit shown in figure 39-1, the 741 op amp will be used as an inverted amplifier to detect when one voltage becomes greater than another.

USING THE OP AMP AS AN INVERTED AMPLIFIER

Notice that this circuit does not use an above- and below-ground power supply. Instead, it is connected to a power supply with a single positive and negative output. During normal operation, the noninverting input of the amplifier is connected to a zener diode. The zener diode produces a constant positive voltage at the noninverting input of the amplifier and is used as a reference. As long as the noninverting input is more positive than the inverting input, the output of the amplifier will be high. A light-emitting diode (LED), D1, is used to detect a change in the polarity of the output. As long as the output of the op amp remains high, the LED stays turned off. This is because the LED has equal voltage applied to both its anode and cathode. Since both are connected to +12 volts, there is no potential difference. Therefore, no current flows through the LED.

If the voltage at the inverting input becomes more positive than the reference voltage applied to pin #3, the output voltage will go

OBJECTIVES

After studying this unit the student should be able to:

- Describe the operation of a level detector
- Connect an operational amplifier as a noninverting level detector
- Connect an operational amplifier as an inverting level detector

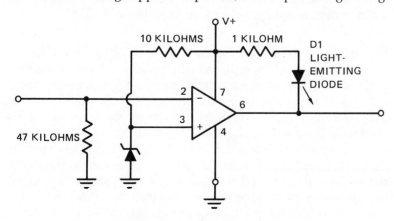

Figure 39-1 Inverting level detector

low, to about +2.5 volts. The output voltage of the op amp will not go to zero or ground because it is not connected to a voltage which is below ground. If the output voltage is to go to 0 volt, pin #4 must be connected to a voltage which is below ground. When the output goes low there is a potential of about 9.5 volts (12 − 2.5 = 9.5) produced across R1 and D1. This causes the LED to turn on and indicate that the state of the op amp's output has changed from high to low.

AS A DIGITAL DEVICE?

In this circuit, the op amp appears to be a digital device because the output seems to have only two states, high or low. Actually, it is not a digital device: this circuit only makes it appear to be digital. Notice in figure 39-1 that there is no negative feedback loop connected between the output and the inverting input. The amplifier uses its open loop gain (which is about 200,000 for the 741), to amplify the voltage difference between the inverting and the non-inverting input. If:

1. the voltage applied to the inverting input becomes one millivolt more positive than the reference voltage applied to the noninverting input,
2. the amplifier will try to produce an output which is 200 volts more negative than its high state voltage (.001 × 200,000 = 200).

The output voltage of the amplifier cannot be driven 200 volts more negative, of course. Because there are only 12 volts applied to the circuit, the output voltage simply reaches the lowest voltage it can and then goes into saturation. As shown, the op amp is not a digital device, but it can be made to act like one.

Replacing the Zener with a Voltage Divider

If the zener diode is replaced with a voltage divider, as shown in figure 39-2, the reference voltage can be set to any value desired. By adjusting the variable resistor in the figure, the positive voltage applied to the noninverting input can be set the same way. If the voltage at the noninverting input is set for 3 volts, the output of the op amp will go low when the voltage applied to the inverting input becomes greater than +3 volts. If the voltage connected to the noninverting input is changed to +8 volts, the output voltage will go low when the inverting input voltage becomes greater than +8 volts. This circuit permits the voltage level at which the output of the op amp will change to be adjusted.

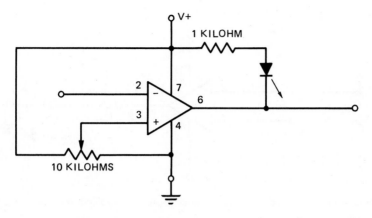

Figure 39-2 Adjustable inverting level detector

CHANGING THE OUTPUT
FROM LOW TO HIGH

In these two circuits, the op amp changed from a high level to a low level when activated. There may be occasions, however, when it is desirable to change the output from a low to a high level. This can be done by connecting the inverting input to the reference voltage and the noninverting input to the voltage being sensed, figure 39-3.

In this circuit, the zener diode provides a positive reference voltage to the inverting input. As long as this voltage remains more positive than the voltage at the noninverting input, the output voltage of the op amp will remain low. If the voltage applied to the noninverting input becomes more positive than the reference voltage, the output of the op amp will become high.

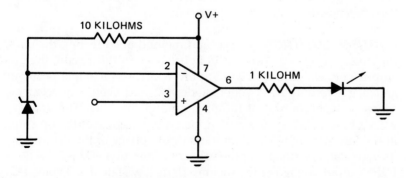

Figure 39-3 Noninverting level detector

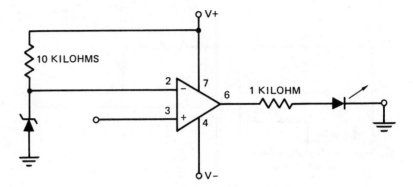

Figure 39-4 Below-ground power connection permits the output voltage to become negative.

A Minor Problem with the Circuit

Depending on the application, this circuit could cause a minor problem. Since this circuit does not use an above- and below-ground power supply, the low output voltage of the op amp will be about +2.5 volts. This positive voltage could cause any devices connected to the output to turn on even if they should be off. If the LED in figure 39-3 is used, it will glow dimly when the output is in the low state. This problem can be corrected in two ways.

A SOLUTION: One way is to connect the op amp to an above- and below-ground power supply as shown in figure 39-4. The output voltage of the op amp in this circuit is negative (below ground) as long as the voltage applied to the inverting input is more positive than the voltage applied to the noninverting input. As long as the output voltage of the op amp is negative with respect to ground, the LED is reverse biased and cannot operate. When the voltage applied to the noninverting input becomes more positive than the voltage applied to the inverting input, the output of the op amp becomes positive and the LED turns on.

ANOTHER SOLUTION: The second method of correcting the output voltage problem is shown in figure 39-5. In this circuit, the op amp is connected to a power supply which has a single positive and negative output. A zener diode, D2, has been connected in series with the output of the op amp and the LED. The voltage value of diode D2 is greater than the output voltage of the op amp in the low state, but less than the output voltage of the op amp in its high state. Assume that the value of zener diode D2 is 5.1 volts. If the output voltage of the op amp in its low state is 2.5 volts, D2 is turned off and will not conduct. If the output voltage becomes

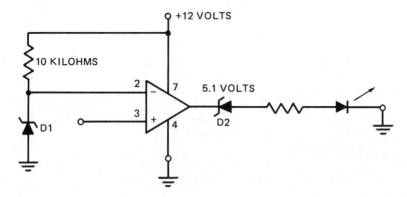

Figure 39-5 A zener diode is used to keep the output turned off.

+12 volts (in the op amp's high state), the zener diode turns on and conducts current to the LED. The zener diode, D2, keeps the LED turned completely off until the op amp switches to its high state. It then provides enough voltage to overcome the reverse voltage drop of the zener diode.

SUMMARY

In the preceding circuits, an LED was used to indicate the output state of the amplifier. Keep in mind that the LED is used only as a detector, and that the output of the op amp can be used to control almost anything. For example, the output of the op amp can be connected to the base of a transistor as shown in figure 39-6. The transistor can then control the coil of a relay. The relay can be used to control almost anything.

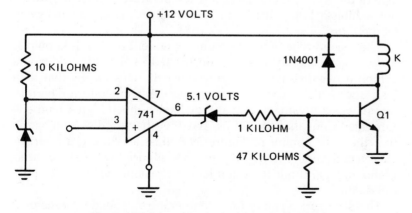

Figure 39-6 A transistor is used to control a relay.

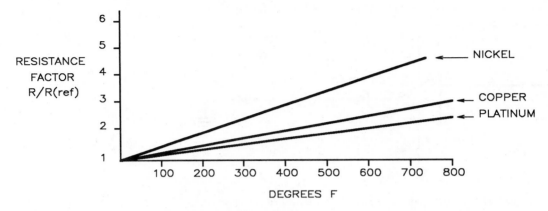

Figure 39-7 Characteristic curves of different RTDs

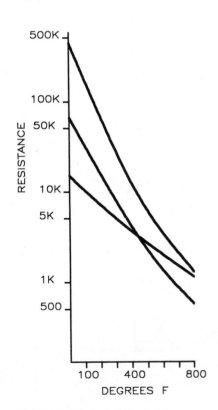

Figure 39-8 Typical characteristic curves of thermistors

TEMPERATURE DEPENDENT RESISTORS

In this circuit, the temperature of the water in the tank, and the temperature of the collector are sensed by temperature dependent resistors. Temperature dependent resistors are devices that exhibit a change of resistance with a change of temperature. There are two basic types of temperature dependent resistors. One type is called a *resistive temperature detector* or *RTD*. RTDs are constructed of metal and have a *positive temperature coefficient*. The temperature coefficient is the ratio of change in resistance as compared to a change in temperature. The word positive means that the resistance of the device will increase as the temperature increases. The temperature coefficient of an RTD is determined by the metal of which it is constructed. As a general rule, RTDs will have a linear coefficient of temperature. The characteristic curve of an RTD will approximate a straight line as shown in figure 39-7. The disadvantage of the RTD is that it has a low temperature coefficient. It does not exhibit a large change in resistance over a large change of temperature.

The second type of temperature dependent resistor is known as a *thermistor*. Thermistors are constructed of metallic oxides and exhibit a negative temperature coefficient. This means that their resistance will decrease with an increase of temperature. Thermistors have an advantage in that they have a much higher temperature coefficient than do RTDs, but they have a disadvantage in that they are nonlinear. Figure 39-8 illustrates the characteristic curves of several different thermistors. Notice the resistance scale is shown logarithmic to permit higher resistance values to be indicated.

The schematic symbol for a temperature dependent resistor is generally a resistor symbol drawn inside a circle with the letter T written beside it. An arrow through the resistor indicates the resistor is variable, figure 39-9. Some schematic diagrams point the arrow toward the top of the circle to indicate the use of an RTD, and toward the bottom of the circle to indicate the use of a thermistor. This, however, is not a universally accepted practice.

In the circuit shown in figure 39-10, the operational amplifier is connected to +12 volts at pin #7 and pin #4 is at ground potential. The output of the op amp is intended to operate the input of a solid state relay, which in turn controls the operation of the pump motor. Since pin #4 has been connected to ground instead of a negative voltage, a 5.1 volt zener diode has been connected between the output of the op amp and the input of the solid-state relay. The zener diode prevents the relay from being turned on when the op amp is in the low state.

Thermistor T_2 and resistor R_3 form a voltage divider circuit for the noninverting input. Thermistor T_2 is placed in a position that will permit it to sense the temperature of the water in the storage tank. Thermistor T_1 and resistors R_1 and R_2 form a voltage divider circuit for the inverting input. Thermistor T_1 is attached to the collector, which permits it to sense the collector's surface temperature. Resistor R_2 is variable to permit a range of adjustment. It is the setting of this resistor that will determine the difference in temperature that is necessary between the collector and the tank to turn on the pump.

To understand the operation of the circuit, assume that the water in the tank is at a higher temperature than the surface of the

RTD   THERMISTOR

Figure 39-9 Schematic symbols for temperature dependent resistors

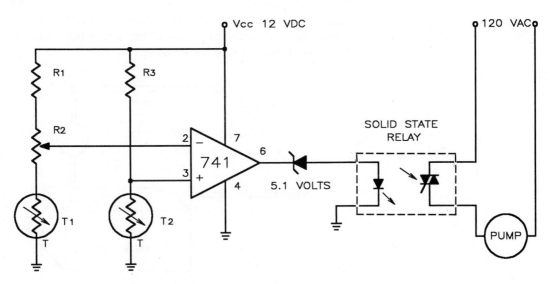

Figure 39-10 Circulating pump control for a solar heating system

collector. Resistor R_2 has been adjusted to permit the voltage applied to the inverting input to be higher than the voltage applied to the noninverting input. The output of the op amp is low and the pump motor is turned off. Now assume that the temperature of the collector begins to increase. This increase of temperature causes the resistance of thermistor T_1 to decrease. As the resistance of T_1 decreases, the voltage applied to the inverting input decreases also. When the voltage applied to the inverting input becomes lower than the voltage applied to the noninverting input, the output of the op amp turns on and starts the pump motor.

As water is calculated from the storage tank to the collector, it will increase in temperature. The increased water temperature will cause the resistance of T_2 to decrease and lower the voltage applied to the noninverting input. When the water has been heated to within a few degrees of the collector temperature, the resistance of T_2 becomes low enough to permit the output of the op amp to turn off.

APPLICATIONS

A good example of how an operational amplifier can be used as a level detector can be seen in figure 39-10. This circuit is used to control the operation of a pump motor that circulates water from a storage tank to a solar collector, as shown in figure 39-11. Pump operation is determined by the difference in temperature between the water in the tank and the collector. If the water in the tank is at a higher temperature than the surface of the collector, the pump motor is turned off. When the collector temperature becomes greater than the tank temperature, the pump turns on and circulates water from the tank to the collector.

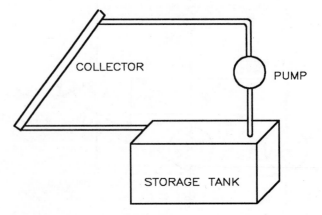

Figure 39-11 Pump circulates water from tank to collector.

UNIT 40

The 741 Operational Amplifier Used as an Oscillator

An operational amplifier can be used as an oscillator. The circuit shown in figure 40-1 is a very simple circuit which produces a square wave output.

THE CIRCUIT

This circuit is rather impractical, however. The circuit depends on a slight imbalance in the op amp or random circuit noise to start the oscillator. Recall from previous units that a voltage difference of only a few millivolts is all that is needed to make the output of the amplifier go high or low. For example, if the inverting input becomes slightly more positive than the noninverting input, the output goes low (negative). With the output negative, capacitor Ct charges through resistor Rt to the negative value of the output voltage. When the voltage applied to the inverting input becomes slightly more negative than the noninverting input, the output changes to a high (positive) value of voltage. With the output positive, capacitor Ct charges through resistor Rt toward the positive output voltage. This circuit works quite well if:

1. the op amp has no imbalance, and
2. the op amp is shielded from all electrical noise.

In practical application, however, there is generally enough imbalance in the amplifier or enough electrical noise to send the op amp into saturation. This stops the operation of the circuit.

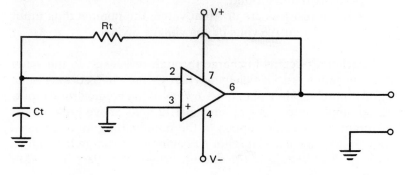

Figure 40-1 Simple square-wave oscillator

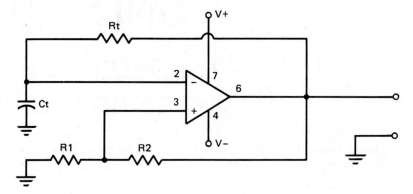

Figure 40-2 Square-wave oscillator using a hysteresis loop

Adding a Hysteresis Loop

The real problem with this circuit is that a millivolt difference between the two inputs is enough to drive the amplifier's output from one state to the other. This can be corrected with the addition of a hysteresis loop connected to the noninverting input as shown in figure 40-2.

OPERATION OF THE CIRCUIT

Resistors R1 and R2 form a voltage divider for the noninverting input. They are generally of equal value. To understand the circuit operation:

1. Assume that the inverting input is slightly more positive than the noninverting input.
   - This causes the output voltage to go negative.
2. Also assume that the output voltage is now negative 12 volts as compared to ground.
   - If the resistors are of equal value, the noninverting input is driven to −6 volts by the voltage divider.

Capacitor Ct begins to charge through resistor Rt to the value of the negative output voltage. When Ct has charged to a value slightly more negative than the −6 volts at the noninverting input, the op amp's output goes high (to +12 volts above ground). At this point, the voltage applied to the noninverting input changes from −6 to +6 volts. Capacitor Ct begins to discharge through Rt to the positive voltage of the output. When the voltage applied to the inverting input becomes more positive than the noninverting input, the output changes to a low value (− 12 volts). The voltage at the noninverting input is driven from +6 to −6 volts, and capacitor Ct again begins to charge toward the negative output voltage.

The addition of the hysteresis loop greatly changes the operation of the circuit. The differential between the two inputs is now volts instead of millivolts. The output frequency of the oscillator is determined by the values of Ct and Rt. The period of one cycle can be computed by using the formula T = 2RC.

THE OP AMP AS A PULSE GENERATOR

The operational amplifier can also be used as a pulse generator. The difference between an oscillator and a pulse generator is the period of time that the output remains on as compared to the time it remains low or off.

An oscillator produces a waveform which has positive and negative pulses of equal voltage and time. In figure 40-3, notice that the positive and negative values of voltage are the same. Both the positive and negative cycles remain turned on for the same

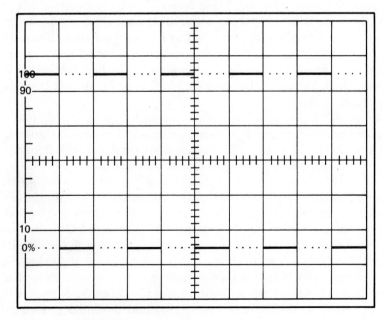

Figure 40-3 Output of an oscillator

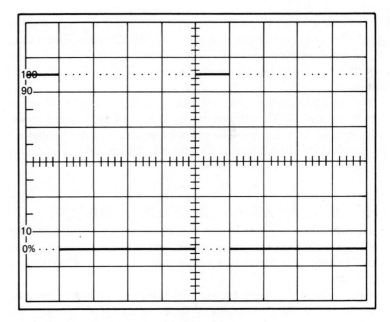

Figure 40-4 Output of a pulse generator

amount of time. This waveform is the same as one seen when an oscilloscope is connected to the output of a square wave oscillator.

If the oscilloscope is connected to a pulse generator, however, a waveform similar to the one shown in figure 40-4 will be seen. The positive value of voltage is the same as the negative just as in figure 40-3. However, the positive pulse is much shorter than the negative pulse.

The 741 operational amplifier can easily be changed from a square wave oscillator to a pulse generator. The circuit shown in figure 40-5 is the same as the square wave oscillator with the addition of resistor R3 and R4, and diodes D1 and D2.

It permits capacitor Ct to charge at a different rate when the output is high (positive) than when the output is low (negative). Assume that the voltage of the op amp's output is low (− 12 volts). Since the output voltage is negative, D1 is reverse biased and no current flows through R3. Therefore, Ct must charge through R4 and D2 which is forward biased. When the voltage applied to the inverting input becomes more negative than the voltage applied to the noninverting input, the output of the op amp becomes + 12 volts. Now diode D2 is reverse biased and diode D1 is forward biased. Capacitor Ct begins charging toward the + 12 volts through resistor R3 and diode D1.

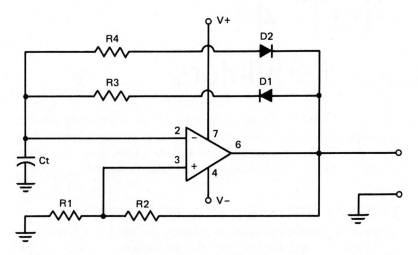

Figure 40-5 Pulse generator circuit

Notice that the amount of time the output of the op amp remains low is determined by the value of Ct and R4. The amount of time the output remains high is determined by the value of Ct and R3. These two ratios can be determined by the ratio of resistor R3 to resistor R4.

UNIT 41

Voltage Regulators

OBJECTIVES

After studying this unit the student should be able to:

- Describe the operation of a series regulator
- Describe the operation of a shunt regulator
- Discuss the need for voltage regulation in a circuit
- Discuss current limit control
- Construct a voltage regulated power supply with current limit

A device used to change ac into dc voltage is generally referred to as a power supply. They range in complexity from a simple half-wave rectifier as shown in figure 41-1 to a unit which is voltage regulated, current limited, and temperature protected.

THE BATTERY AS A POWER SUPPLY

Many of the circuits used in industry are sensitive to a change in voltage. A regulator must be used to provide a constant output voltage. The voltage of an unregulated power supply changes with a change in load current. This change is caused by the internal impedance of the circuit. A nearly perfect dc power supply is probably the battery, but it, like all components, has some internal impedance. It is generally very low, but nevertheless, it's there.

Assume a 12-volt battery has an internal impedance of .1 ohm, figure 41-2. If a 1.2-ohm resistor is connected to the battery terminals, 10 amps of current will attempt to flow through the load. This current causes a voltage drop across the terminals of 1 volt (10 × .1 = 1). Notice that even a battery has some voltage drop when a load is connected to it. If the battery did not have internal impedance, it could produce unlimited current (12/0 = infinity). Electronic power supplies are similar because they have internal impedance too.

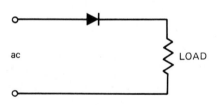

Figure 41-1 Half-wave rectifier

Adding a Variable Resistor

This problem can be corrected, however. Assume that a dc power supply is needed which can furnish 2 amps of current at 12 volts. To do this a power supply must first be constructed which can furnish more than 12 volts at 2 amps; for example, 14 volts at

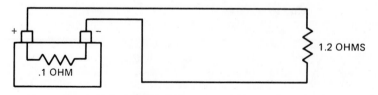

Figure 41-2 Internal impedance

184

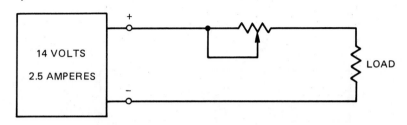

Figure 41-3 Series resistor used to regulate voltage to the load

2.5 amps. A variable resistor connected in series with the load can adjust the voltage at the output if the load current should change, figure 41-3.

If the load connected to the output is small (.1 amp), resistance can be added in series to produce a voltage of 12 volts at the output. If the current is increased to 1.5 amps, the resistance can be decreased to produce 12 volts. Notice that the output voltage of the power supply can be adjusted (regulated) by the amount of series resistance in the circuit. This type of regulator isn't too practical, however, because:

1. it requires the constant attention of an operator, and
2. people don't have reflexes that are fast enough to catch a sudden increase or decrease in load current.

THE ZENER DIODE AS A REGULATOR

The simplest electronic voltage regulator is the zener diode, which was covered in unit 9. Almost all solid-state devices have a characteristic known as *dynamic* impedance: meaning changing.

A common junction diode has a forward voltage drop of .6 to .7 volt regardless of the amount of current flowing through it. In order for the drop to remain constant, the impedance of the device must change when current changes. Assume that a diode has a forward current of .05 amp. Its impedance is 12 ohms (.6/.05 = 12). If the current is increased to 1 amp, the impedance now appears to be .6 ohm (.6/1 = .6).

The zener diode acts the same when connected in the reverse direction, figure 41-4. Regardless of the current flowing through the zener in the reverse direction, its voltage drop remains constant.

The zener diode is generally used as a shunt regulator; it is connected in parallel with the load. The circuit shown in figure 41-5 is generally used for low power applications. Power zeners

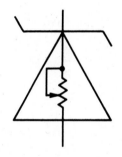

Figure 41-4 Dynamic impedance

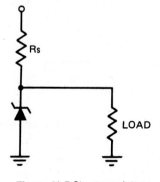

Figure 41-5 Shunt regulator

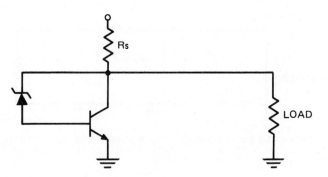

Figure 41-6 Power zener circuit

are available if more power is needed. A power transistor controlled by a zener diode can be used to handle the power of the circuit, figure 41-6.

Disadvantages of the Zener Diode

Shunt regulators are used in some applications, but they have some disadvantages. They require the use of a series resistor, Rs. This limits the current flow through the transistor when no load is connected to the output. It also adds internal impedance to the power supply which puts a definite limit on the output current.

SERIES CONTROL AS A SOLUTION: Most voltage regulators use a series control as shown in figure 41-7. This type of regulator controls the output voltage by either

1. turning the transistor on harder if the voltage tries to drop, or
2. turning it off if the voltage tries to increase.

The zener diode connected between base and ground tries to keep the emitter voltage at the same potential as the base. If a load is connected between the emitter and ground, the voltage at the emitter will try to drop. When this happens, the zener diode begins to raise its internal impedance. This causes more base current to flow to the transistor. The increase of base current causes the transistor to turn on harder and raises the voltage of the emitter back to the voltage of the zener diode.

Although the circuit shown in figure 41-7 works great on paper, it has a problem: the percentage of regulation is proportional to the amount of gain in the circuit, and power transistors are not famous for their gain.

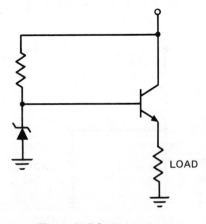

Figure 41-7 Series regulator

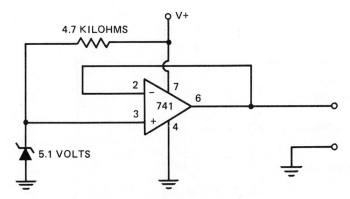

Figure 41-8 Simple op amp regulator

THE OP AMP AS A VOLTAGE REGULATOR

Operational amplifiers, however, are famous for their gain. The op amp can be used to make a voltage regulator with a high percentage of regulation.

The operational amplifier shown in figure 41-8 uses a zener diode to provide a reference voltage to the noninverting input. The inverting input is connected to the output of the op amp. Connecting the inverting input directly to the output forces the output to assume the same voltage as that applied to the noninverting input. Remember: *The voltages applied to the inverting and noninverting inputs must be equal.*

Testing the Op Amp

One of the easiest tests to perform on op amps in a working circuit is to measure the voltage drop between the inverting and noninverting inputs with a high impedance voltmeter. If the voltage is not zero or close to zero, the op amp is probably defective. Therefore, if 5.1 volts is applied to the noninverting input, the inverting input must be 5.1 volts also. Since the inverting input is connected directly to the output, the output voltage must remain at 5.1 volts.

Changing the Output Voltage

The output voltage can be changed by connecting the inverting input to a voltage divider as shown in figure 41-9. In this circuit, the inverting input is connected to a voltage divider made with two resistors of equal value. The inverting input must assume the same voltage as the noninverting input. Therefore, the output

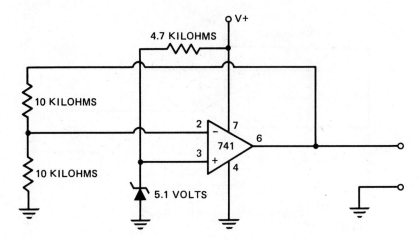

Figure 41-9 Negative feedback can be used to change the output voltage.

voltage of the op amp must become 10.2 volts to produce 5.1 volts at the inverting input. The output voltage of the regulator can be easily controlled by a simple voltage divider. If the voltage divider is replaced by a potentiometer as in figure 41-10, the output voltage can be adjusted to any level between the reference voltage and the maximum output voltage of the op amp.

ADJUSTING THE REGULATOR

The regulator shown in figure 41-10 still has a problem. If a regulator is needed which is adjustable, the output voltage must be adjustable to below 5.1 volts. This is easily done by using a very

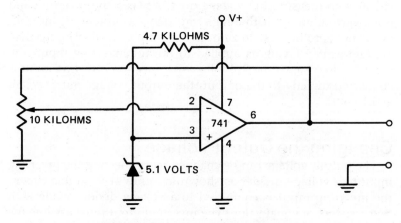

Figure 41-10 Adjustable output voltage

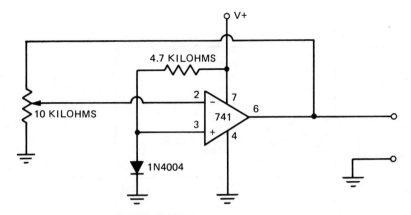

Figure 41-11 Junction diode used as a reference

low voltage zener diode which permits adjustment of the voltage to a low level. Remember that a common junction diode has a forward voltage drop of about .6 volt regardless of the current flow through it. The forward voltage drop of a junction diode, therefore, can be used as a zener of .6 volt, figure 41-11. This circuit permits the output voltage to be adjusted to its lowest level (about 2 volts). Recall that when the V− connection at pin #4 is connected to ground instead of a voltage which is below ground, the output voltage cannot drop below about 2 volts.

This condition can be corrected by connecting the op amp to an above- and below-ground power supply as shown in figure 41-12. Connecting to a below-ground voltage means that the output voltage can now be adjusted from .6 volt to the maximum the op amp can deliver.

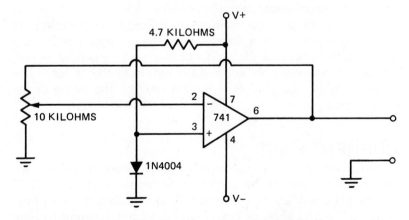

Figure 41-12 Below-ground connection permits the output voltage to be adjusted to a low value.

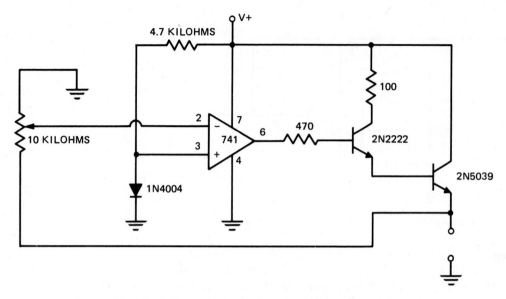

Figure 41-13 Darlington amplifier circuit used to drive a power transformer

Adding a Darlington Driver Circuit

The voltage regulator in figure 41-12 has excellent regulation and is adjustable from .6 volt to the full output voltage of the op amp. It doesn't, however, have the ability to produce much current. The 741 operational amplifier has a maximum output current of about 5 milliamperes (ma). This 5 ma can be used to control the base of a transistor connected in series with the load. To improve circuit operation, a Darlington driver circuit is generally used to provide more transistor gain, figure 41-13.

In this circuit, the potentiometer is connected to the emitter of the power transistor, not the output of the op amp. The connection is changed because the emitter of the transistor is now the output of the power supply and not the output of the op amp. The output voltage and current is limited only by the rating of the components.

CURRENT LIMIT

Many power supplies use a circuit which limits the output current to a safe level. If the output of the regulator in figure 41-13 becomes shorted, enough current will flow to destroy the components in the circuit. Some power supplies have a current-limiting circuit. This permits adjustment of the maximum output current the regulator can produce. A simpler circuit can be used to limit the output current to a safe value. This is generally called short-circuit protection. Most current-limiting circuits used in dc power supplies operate by sensing the voltage drop across a low value series resistor, figure 41-14.

Voltage Drop

If a load is connected to the output terminals of the power supply, current flows through the series resistor. The voltage dropped across this resistor is proportional to the current flow. For instance, if 1 amp of current flows through the resistor, a drop of 1 volt is developed across it. Since this drop is proportional to the current flow, a circuit can be built that does two things:

1. it senses the voltage drop and
2. turns the transistor off when the current becomes excessive.

This same principle is used in small power supplies designed to operate electronic experiments to large SCR controlled power supplies designed to provide the power needed to drive large dc motors. A common sensing resistor found in large power supplies designed to produce several hundred amps is a precision .1-ohm resistor rated at several hundred watts. In the power supply shown in figure 41-15, a 1-ohm wire-wound resistor will be used.

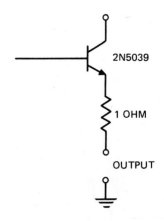

Figure 41-14 One ohm current-sensing resistor

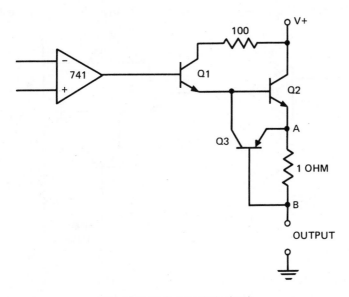

Figure 41-15 Current limit circuit

Adding Short-circuit Protection

The short-circuit protection circuit is very simple to construct, figure 41-15. Transistors Q1 and Q2 form the Darlington driver pair which controls the output voltage. Transistor Q3 provides short-circuit protection. When current flows through the 1-ohm sensing resistor, a voltage drop is produced across it. Point A becomes more positive than point B. If Q3 is a silicon PNP transistor, it begins to conduct when the voltage between its emitter and base reaches about .6 volt. When Q3 begins turning on, it steals the base current from Q2 and causes it to begin turning off.

Notice that transistor Q3 will not permit a drop of more than about .6 volt to exist between its base and emitter connections. The current through the sense resistor does not become greater than that necessary to produce a drop of about .6 volt. In this circuit the current is limited to about .6 amp ($1 \times .6 = .6$).

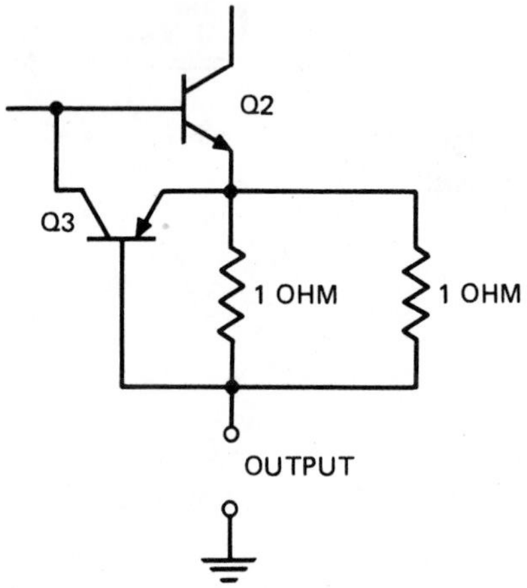

Changing the Current Limit

If a different current limit is desired, two changes can be made. One is to change the value of the sense resistor. For example, if another 1-ohm resistor is connected in parallel with the existing one, as shown in figure 41-16, the total value of resistance is .5 ohm ($1/2 = .5$). This circuit requires a flow of about 1.2 amps to produce a drop of .6 volt across the emitter-base junction of Q3.

Another method is to increase the voltage drop needed to turn transistor Q3 on. This is done by connecting a junction diode in series with the base of transistor Q3, figure 41-17. A diode con-

Figure 41-16 Current-sensing resistor is decreased to ½ ohm

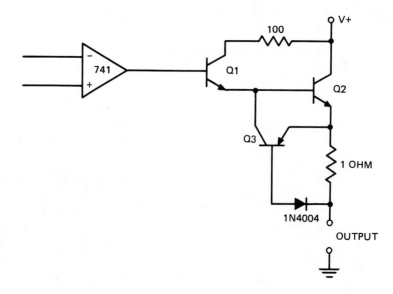

Figure 41-17 Voltage drop of the current-limit circuit is increased to about 1.2 volts

nected in the forward bias direction has a voltage drop of .6 volt. Therefore:

- .6 volt drop of the emitter-base junction of Q3 plus
- .6 volt drop of diode D1 requires a drop across the sense resistor of
- 1.2 volts to turn Q3 on.

Of the two methods just described, the best is to lower the value of the sense resistor. This method places less resistance in series with the output of the power supply.

UNIT 42
Digital Logic

OBJECTIVES

After studying this unit the student should be able to:

- List different types of logic
- Discuss the operation of AND, OR, NAND, NOR, and INVERTER gates
- Draw both USASI and NEMA logic symbols
- Discuss the use of truth tables for different types of logic gates
- Connect logic gates in a circuit and produce a truth table for each type of gate

The electrician in industry today must be familiar with solid-state digital-logic circuits. An example of these digital logic, or digital integrated circuits, is shown in figure 42-1, attached to a heat sink. A digital device is one which has only two states, on or off. Most electricians have been using them for many years without realizing it. Magnetic relays, for instance, are digital devices. The coil is the input and the contacts are the output. They are considered to be single-input multi-output devices, figure 42-2.

Although relays are digital devices, the term *digital logic* has come to mean circuits that use solid-state control devices known as gates. The five basic types of gates are: the AND, OR, NOR, NAND, and INVERTER or NOT. Each of these gates will be covered later in the unit.

VARIOUS TYPES OF DIGITAL LOGIC

There are also different types of logic. One of the earliest to appear was RTL, or resistor transistor logic. Next was DTL, or diode transistor logic, and then TTL, or transistor transistor logic. RTL and DTL have about faded out of existence, but TTL is still used to a fairly large extent. TTL can be identified because it operates on 5 volts.

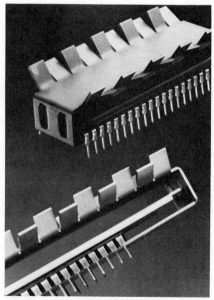

Figure 42-1 Inline integrated circuit attached to a heat sink

PHOTO COURTESY OF AAVID ENGINEERING CO.

Figure 42-2 Magnetic relay

PHOTO COURTESY OF EATON CORP., CUTLER-HAMMER PRODUCTS

Another type of logic frequently used in industry is HTL, which stands for high transit logic. It is better than TTL at ignoring the voltage spikes and drops caused by the starting and stopping of inductive devices such as motors. HTL generally operates on 15 volts.

Another very popular type of logic is CMOS, which has very high input impedance. CMOS is shortened from COSMOS, which means complimentary symmetry metal oxide semiconductor. CMOS logic has the advantage of requiring almost no power to operate. There are also disadvantages:

- CMOS logic is very sensitive to voltage, and
- the static charge in a body can sometimes destroy an integrated circuit (IC) just by touching the circuit.

People who work with CMOS logic often use a ground strap which is worn around the wrist like a bracelet. The strap is used to prevent a static charge from building up on the body.

A characteristic of CMOS logic is that unused inputs cannot be left in an indeterminent state. They must be connected to either a high or a low state.

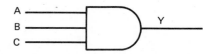

Figure 42-3 USASI symbol of a three-input AND gate

THE *AND* GATE

Whereas magnetic relays are single-input, multi-output devices, gate circuits are multi input, single output. For instance, an AND gate has at least two inputs, but only one output. Figure 42-3 shows an AND gate with three inputs, labeled A, B, and C, and one output labeled Y.

The AND gate symbol shown in figure 42-3 is a USASI standard logic symbol. These symbols are commonly referred to as computer logic symbols. There is another system known as NEMA logic which uses completely different symbols. The NEMA symbol for a three-input AND gate is shown in figure 42-4.

Both symbols mean the same although they are drawn differently. Electricians in industry must learn both sets, since both types are used. Regardless of which symbol is used, the AND gate operates the same way.

The rule for an AND gate is that it must have all of its inputs high in order to get an output. Assuming that TTL logic is being used, a high level is considered to be +5 volts and low to be 0 volt.

The truth table shown in figure 42-5 illustrates the state of a gate's output with different conditions of input. Ones are used to represent a high state and zeros a low state. Notice in the figure

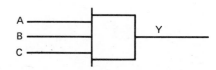

Figure 42-4 NEMA logic symbol of a three-input AND gate

A	B	Y
0	0	0
0	1	0
1	0	0
1	1	1

Figure 42-5 Truth table for a two-input AND gate

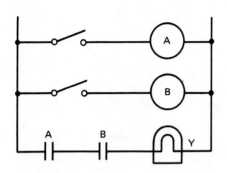

Figure 42-6 Relay equivalent circuit of a
two-input AND gate

A	B	C	Y
0	0	0	0
0	0	1	0
0	1	0	0
0	1	1	0
1	0	0	0
1	0	1	0
1	1	0	0
1	1	1	1

Figure 42-7 Truth table for a three-input AND gate

that the output of the AND gate is high only when both of its inputs are high. The AND gate operates much like the simple relay circuit shown in figure 42-6.

If a lamp is used to indicate the output of the AND gate, both A and B must be energized before there is an output. Figure 42-7 shows the truth table for a three-input AND gate. There is still only one condition that permits a high output for the gate: when all inputs are high (at a logic level 1). When using an AND gate, remember: *any 0 input = a 0 output.* An equivalent relay circuit for a three-input AND gate is shown in figure 42-8.

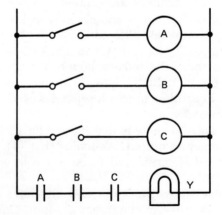

Figure 42-8 Relay equivalent circuit of a three-input AND gate

Figure 42-9 (A) Computer logic symbol for an OR gate (B) NEMA logic symbol for an OR gate

A	B	Y
0	0	0
0	1	1
1	0	1
1	1	1

Figure 42-10 Truth table for a two-input OR gate

THE *OR* GATE

The next gate to study is the OR gate. Both the computer (USASI) and the NEMA logic symbols are shown in figure 42-9.

The OR gate has a high output when either or both inputs are high: *any 1 input = a 1 output*. The truth table is shown in figure 42-10. An equivalent relay circuit for the OR gate is shown in figure 42-11. Notice in this circuit that if either or both of the relays are energized, there will be an output at Y.

THE *EXCLUSIVE-OR* GATE

Another gate that is very similar to an OR gate is known as an EXCLUSIVE-OR gate. The symbol for this gate is shown in figure 42-12.

The EXCLUSIVE-OR gate has a high output when either but

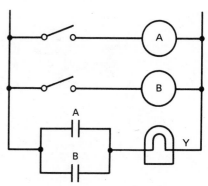

Figure 42-11 Relay equivalent circuit of a two-input OR gate

Figure 42-12 Computer logic symbol of an EXCLUSIVE-OR gate

A	B	Y
0	0	0
0	1	1
1	0	1
1	1	0

Figure 42-13 Truth table for a two-input EXCLUSIVE-OR gate

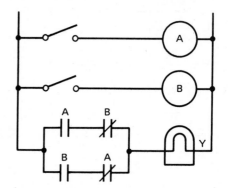

Figure 42-14 Relay equivalent circuit of a two-input EXCLUSIVE-OR gate

not both inputs are high. Refer to the truth table for this gate in figure 42-13. Notice that the equivalent relay circuit in figure 42-14 shows that if both are deenergized or if both are energized there is no output.

THE *INVERTER* (NOT) GATE

The simplest gate is the INVERTER or NOT gate. The IN-VERTER has one input and one output. As the name implies, the output is inverted (opposite the input). For example, if the input is high the output will be low, and if the input is low the output will be high. Figure 42-15 shows the computer and NEMA symbols for the INVERTER. Notice that the computer symbol is an ampli-

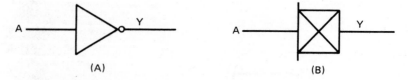

Figure 42-15 (A) Computer logic symbol of an INVERTER (NOT) (B) NEMA logic symbol of an INVERTOR (NOT)

fier with a small circle (O), drawn at the output point. When using computer logic, an O drawn on a gate means to invert. Since the O appears on the output end of the amplifier, it means that the output is inverted. NEMA symbols use an X to show this. The truth table in figure 42-16 clearly shows that the output of the INVERTER is opposite the input. Figure 42-17 is an equivalent relay circuit for the INVERTER.

THE *NOR* GATE

The next gate to study is the NOR gate. The word NOR is shortened from NOT OR. The computer and NEMA logic symbols for a NOR gate are shown in figure 42-18.

Notice that the NOR gate symbol is the same as the OR gate symbol with an inverted output. A NOR gate is made by connecting an INVERTER to the output of an OR gate as shown in figure 42-19. The truth table in figure 42-20 shows that the output of a NOR gate is 0 or low when any input is high: *any 1 input = a 0 output.* An equivalent relay circuit for this gate is shown in

A	Y
0	1
1	0

Figure 42-16 Truth table for an INVERTER (NOT)

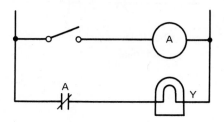

Figure 42-17 Relay equivalent circuit of an INVERTER (NOT)

Figure 42-18 (A) Computer logic symbol of a two-input NOR gate (B) NEMA logic symbol of a two-input NOR gate

Figure 42-19 Equivalent NOR gate

A	B	Y
0	0	1
0	1	0
1	0	0
1	1	0

Figure 42-20 Truth table for a two-input NOR gate

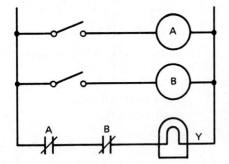

Figure 42-21 Relay equivalent circuit for a two-input NOR gate

figure 42-21. If either relay A or B is energized, there is no output at Y.

THE *NAND* GATE

The last gate to study is the NAND gate. NAND is shortened from NOT AND. Figure 42-22 shows the computer and the NEMA logic symbols for the NAND gate. These symbols are the same as the one for the AND gate with an inverted output. If any input of a NAND gate is low, the output will be high: *any 0 input = 1 output*. The output is 0 only when all inputs are 1. The truth table in figure 42-23 clearly indicates these facts. Figure 42-24 shows an equivalent relay circuit for the NAND gate. If either relay A or B is deenergized, there will be an output at Y.

(A) (B)

Figure 42-22 (A) Computer logic symbol for a two-input NAND gate (B) NEMA logic symbol for a two-input NAND gate

A	B	Y
0	0	1
0	1	1
1	0	1
1	1	0

Figure 42-23 Truth table of a two-input NAND gate

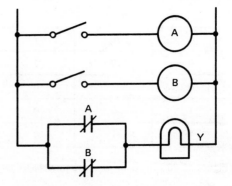

Figure 42-24 Relay equivalent circuit for a two-input NAND gate

Figure 42-25 NAND gate connected as an INVERTER (NOT)

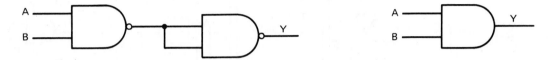

Figure 42-26 NAND gates connected as an AND gate

The NAND gate is referred to as the *basic* gate because it can be used to make any of the other gates. Figure 42-25 shows the NAND gate connected to make an INVERTER. If a NAND gate is used as an INVERTER, and connected to the output of another NAND gate, it becomes an AND gate as shown in figure 42-26.

VARIOUS GATE COMBINATIONS

When two NAND gates are connected as INVERTERS, and these INVERTERS are connected to the inputs of another NAND gate, the OR gate is formed, figure 42-27.

If an INVERTER is added to the output of the OR gate shown in figure 42-27, a NOR gate is formed, figure 42-28.

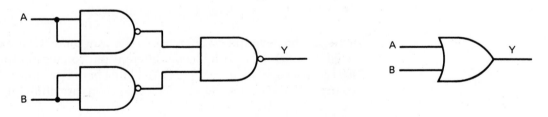

Figure 42-27 NAND gates connected as an OR gate

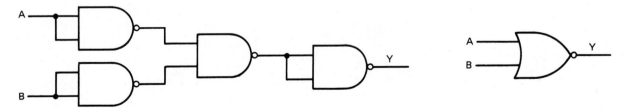

Figure 42-28 NAND gates connected as a NOR gate

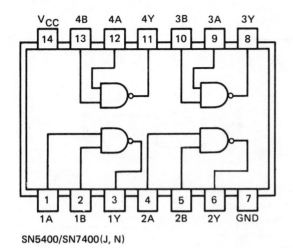

SN5400/SN7400(J, N)

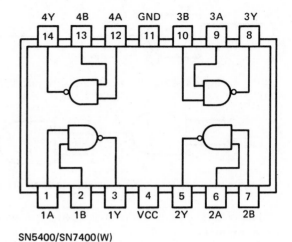

SN5400/SN7400(W)

Figure 42-29 IC connection of a quad two-input NAND gate

COURTESY OF TEXAS INSTRUMENTS INCORPORATED © 1973

HOUSING DIGITAL-LOGIC GATES

Digital-logic gates are generally housed in 14 pin IC packages. One of the types of TTL logic frequently used is the 7400 family of devices. For instance, a 7400 IC is a quad two-input positive NAND gate. This means that:

- There are four NAND gates contained in the package.
- Each gate has two inputs.
- A level 1 is considered to be a positive voltage.

There can, however, be a difference in the way some ICs are connected. A 7400 (J or N) IC has a different pin connection than a 7400 (W) package. Notice in figure 42-29 that both ICs contain four two-input NAND gates, but the pin connections are different.

UNIT 43
The Bounceless Switch

When a control circuit is constructed, it must have sensing devices which tell it what to do. The number and types of devices used are determined by the circuit. They can range from a simple pushbutton to float, limit, and pressure switches. Most of these devices use some type of mechanical switch to indicate their condition.

A float switch, for example, indicates its condition by opening or closing a set of contacts, figure 43-1. The float switch can *tell* the control circuit whether a liquid is at a certain level or not. Most of the other sensing devices use this same method to indicate a condition. A pressure switch indicates that a pressure is at a certain level or not. Limit switches indicate if a device has moved a certain distance or if it is present or absent from some location.

THE SNAP ACTION SWITCH

Almost all of these devices employ a *snap action* switch. The snap action in a mechanical switch is generally obtained by spring-loading the contacts. The snap action ensures good contact when the switch operates.

Assume that a float switch is used to sense when water reaches a certain level in a tank. If the water rises at a slow rate, the contacts come together at a slow rate. This results in a poor connection. If the contacts are spring-loaded, however, they will snap from one position to another when the water reaches a certain level.

Contact Bounce

Most contacts have a snap action but they do not usually close with a single motion. When the movable contacts *make* with the stationary contact, there is often a fast, bouncing action. This means that the contacts may actually make and break three or four times in succession before the switch remains closed. When this switch controls a relay, contact bounce does not cause a problem. This is because relays are relatively slow-acting devices and are not affected, figure 43-2.

When used with electronic control systems, contact bounce can be the source of a great deal of trouble. Most digital-logic circuits are very fast acting. They can count each pulse when a contact

OBJECTIVES

After studying this unit the student should be able to:

- Explain why a bounceless switch is necessary in computer circuits
- Discuss the operation of a bounceless switch
- Describe how a bounceless switch is constructed using logic gates
- Construct a bounceless switch circuit using logic gates

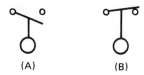

Figure 43-1 (A) Normally open float switch (B) Normally closed float switch

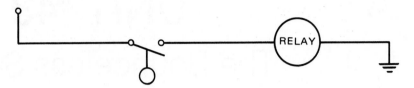

Figure 43-2 Contact bounce does not greatly affect relay circuits.

bounces. Depending on the circuit, it may interpret each of these pulses as a command. Contact bounce can cause the control circuit to literally "lose its mind."

Debouncing the Switch

For this reason, contacts are debounced before they are permitted to talk to the control system. When contacts must be debounced, a circuit commonly called a *bounceless switch* is used. Several circuits can be used to construct a bounceless switch, but the most common method uses digital-logic gates. Any of the inverting gates can be used, but in this unit only two will be used.

OPERATION OF A BOUNCELESS SWITCH

Before construction of the circuit begins, the operation of a bounceless switch should first be discussed. The idea is to construct a circuit which locks its output either high or low when it detects the first pulse from the mechanical switch. If the output is locked in any position, it ignores any other pulses it may receive from the switch. The output of the bounceless switch is connected to the input of the digital-control circuit. The control circuit receives only one instead of a series of pulses.

CONSTRUCTING THE CIRCUIT

The INVERTER

The first gate to be used to construct a bounceless switch is the INVERTER. The computer symbol and the truth table for this gate

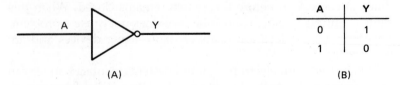

A	Y
0	1
1	0

(A)   (B)

Figure 43-3 (A) INVERTER symbol (B) INVERTER truth table

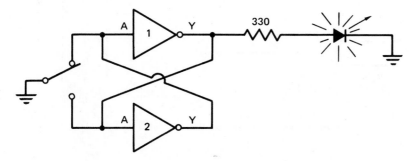

Figure 43-4 High output condition

are shown in figure 43-3. The bounceless switch circuit using INVERTERS is shown in figure 43-4. The output of the circuit is high with the switch in the position shown. The switch connects the input of INVERTER #1 directly to ground or low. This causes the output of INVERTER #1 to be at a high state. This output is connected to the input of INVERTER #2. Since the input of INVERTER #2 is high, its output is low. The output of INVERTER #2 is connected to the input of INVERTER #1. This causes a low condition at the input of INVERTER #1.

If the position of the switch is changed, as shown in figure 43-5, the output will change to low. The switch now connects the input of INVERTER #2 to ground or low. The output is, therefore, high. The high output of INVERTER #2 is connected to the input of INVERTER #1. Since INVERTER #1 has a high connected to its input, its output becomes low. The output of INVERTER #1 is connected to the input of INVERTER #2. This forces a low input to be maintained at INVERTER #2. Notice that the output of one INVERTER is used to lock the input of the other one.

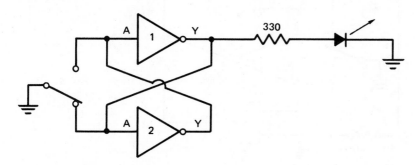

Figure 43-5 Low output condition

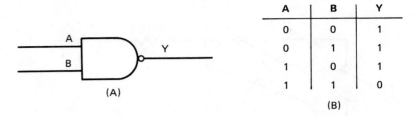

A	B	Y
0	0	1
0	1	1
1	0	1
1	1	0

(B)

Figure 43-6 (A) NAND gate symbol (B) NAND gate truth table

The NAND Gate

The second logic gate to be used to construct a bounceless switch is the NAND gate. The computer logic symbol and the truth table for this gate are shown in figure 43-6. Figure 43-7 shows the construction of a bounceless switch using NAND gates.

In this circuit, the switch has input A of gate #1 connected to low or ground. Since A is low, the output is high. The output of gate #1 is connected to input A of gate #2. Input B of gate #2 is connected to a high through the 4.7-kilohm resistor. Since gate #2 has both inputs high, its output is low. This output is connected to the B input of gate #1. Gate #1 now has a low connected to input B. Its output is forced to remain high even if contact bounce causes a momentary high at input A.

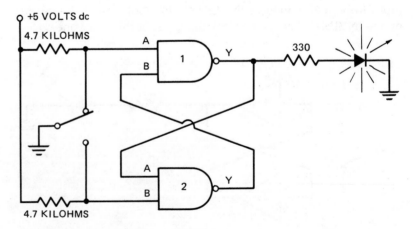

Figure 43-7 High output condition

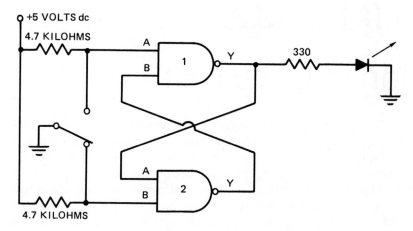

Figure 43-8 Low output condition

When the switch changes position as shown in figure 43-8, input B of gate #2 is connected low. This forces the output of gate #2 to become high. The high output is connected to the B input of gate #1. Input A of gate #1 is connected to high through a 4.7-kilohm resistor. Since both inputs of gate #1 are high, its output is low. The low is connected to the A input of gate #2. This forces its output to remain high even if contact bounce should cause a momentary high at the B input.

The output of this circuit remains constant even if the switch contacts should bounce. The switch is ready to be connected to the input of an electronic control circuit.

UNIT 44
Design of Circuit #1

OBJECTIVES

After studying this unit the student should be able to:

- Discuss the operation of solar cells
- Discuss the operation of a cad cell
- Discuss the design parameters of a photo circuit
- Construct a light-sensitive circuit

The first circuit to be designed is a photodetector. The circuit is used to turn on a light when darkness comes, or activate a control component when a light is either present or absent. First concentrate on turning on a light when it becomes dark.

One of the first things to be learned about circuit design is that there are generally several ways any particular circuit can be constructed and still perform the same function. The final design is usually determined by several factors: the availability of components and their cost. Other factors can be important. These include whether the circuit must be operated in extremes of temperature, or if the circuit must be battery powered.

REQUIREMENTS FOR THE CIRCUIT

One of the first steps in design is to determine what is needed to do the job. This circuit is to be used to turn a light on when it becomes dark and to turn the light off again when it becomes light. It must, therefore, have two basic items:

1. some device to detect the presence or absence of light, and
2. a method to connect the light to the power line.

Two components immediately come to mind: solar cells and a relay.

Solar Cells and the Relay

If enough solar cells are connected to operate a relay, the circuit in figure 44-1 will fulfill the requirements. The light is connected in series with a normally closed relay contact to 120 volts ac. The contact is controlled by CR relay coil, and silicon solar cells are connected to the coil. During the daylight hours the solar cells provide power to the relay coil which keeps the contact open. Since the relay contact is held open during the daylight hours, the light is turned off. During the hours of darkness, the solar cells cannot produce the power needed to operate the relay coil and the contact closes. When it closes, the light is connected to the 120-volt ac power line.

This circuit fulfills all the requirements. A different one may

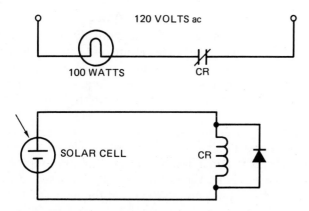

120 VOLTS ac

100 WATTS

CR

SOLAR CELL

CR

Figure 44-1 Photodetector operated by solar cells

have to be used, however, because silicon solar cells are expensive. One cell produces only about .5 volt. If the relay coil operated on 12 volts, it would require 24 solar cells connected in series to produce the 12 volts needed to power the relay coil.

CAD CELLS: Another device which can be used to detect light is the cadmium sulfide cell or *cad cell*, figure 44-2. The cad cell does not produce a voltage in the presence of light as the solar cell does, but rather changes its resistance with a change of light. When it is

Figure 44-2 Cad cells
PHOTO COURTESY OF VACTEC, INC.

exposed to a bright light, its resistance drops to a low value. When it is in darkness, its resistance increases greatly. The exact amount of light or dark resistance varies from one cad cell to another. Typically it has about 50 ohms in direct sunlight and several hundred thousand ohms in darkness. They are economical and can be obtained at almost any electronic supply store.

Connection #1: Examine the circuit shown in figure 44-3. The light is connected in series with a normally closed relay contact. The coil of the 12-volt dc relay is connected in series with a cad cell to the dc power supply. When the cad cell is exposed to sunlight, its resistance drops low enough to permit the relay coil to turn on. When the coil turns on, the normally closed contact opens and disconnects the light from the power source. When the cad cell is in darkness, its resistance increases and turns the relay off. When the relay turns off, the contact closes and connects the light to the ac power source.

This circuit appears to perform all the requirements that were originally set for its operation. There are several problems with it, however. One is that cad cells are very low-power devices and not many are able to control the current through the relay without being destroyed. Another is the lack of control. There is no method that can be used to control *when* the light will turn on or off. A third is that the cad cell will not provide a definite on or off action for the relay. Its resistance changes with a change of light intensity. It does not operate like a digital device which has either a high or a low resistance. Since the cad cell is not a digital device, the current slowly increases through the relay coil as the resistance of the cad cell decreases. This slow increase or decrease of current prevents the relay from operating properly.

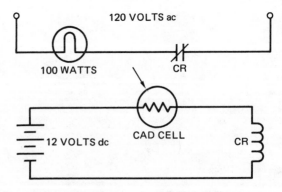

Figure 44-3 Relay coil connected to a cad cell

Connection #2: The circuit shown in figure 44-4 is an improvement over that in figure 44-3. In this circuit, the light is connected in series with a normally-open contact which is controlled by CR relay coil. The coil is connected in series with a transistor which is controlled by the cad cell.

When the cad cell is in the presence of light, its resistance is very low. In this condition it steals the base current to ground and the transistor is turned off, figure 44-5. When the cad cell is in darkness, its resistance increases and current is permitted to flow to the base of the transistor, figure 44-6. When current flows to the base of the transistor, the transistor turns on and energizes the relay. When the relay energizes, CR contacts close and connect the light to the 120-volt ac power source.

This circuit is a definite improvement over that shown in figure 44-3. The transistor of this circuit can easily control the current needed to operate the relay. The variable resistor, R2, can adjust the sensitivity of the circuit. Although this circuit has corrected some of the problems of the circuit in figure 44-3, it still retains one of them.

The transistor is not a digital device and will not turn the coil on or off with a snap action. If this circuit is to operate properly, the photodetector should work with a definite on or off action. A device is required that can convert the analog operation of the cad cell into a digital action.

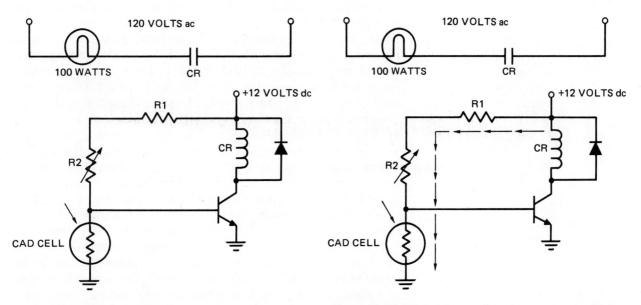

Figure 44-4 Cad cell used to control a transistor   **Figure 44-5** Cad cell resistance is low in the presence of light.

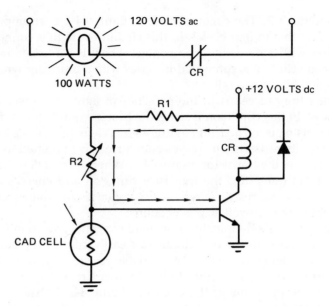

Figure 44-6 Cad cell resistance is high in the presence of darkness.

Adding an Op Amp

The operational amplifier can do this conversion when it is used as a level detector. The circuit in figure 44-7 uses the cad cell to control the output of the operational amplifier. Resistor R1 sets the inverting input to some positive voltage. Assume this voltage level to be 5 volts. Resistor R2 and the cad cell form a voltage-divider circuit for the noninverting input. When the cad cell is in the presence of light its resistance is low. The voltage applied to the non-inverting input is less than 5 volts. Therefore the output of the op amp is low. Since the output is low, the transistor is turned off and CR relay is deenergized. When the cad cell is in darkness, its resistance increases above 4.7 kilohms. The voltage applied to the noninverting input becomes greater than 5 volts. This causes the output of the op amp to change to a high state and supply base current to the transistor. The transistor turns the relay coil on and closes contact CR which connects the light to the power line.

Notice that a circuit now exists that will perform the desired job. It is inexpensive to construct, and the light sensitivity can be adjusted by resistor R1. The op amp permits the cad cell to operate as a digital device. This means that the relay can be turned completely on or completely off.

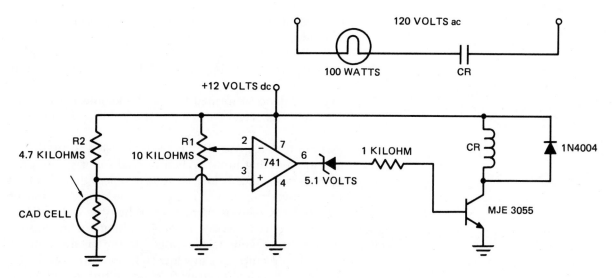

Figure 44-7 An op amp permits the cad cell to operate as a digital device.

CHANGING THE CIRCUIT

Although the circuit design goal has been completed, assume that a change in the sense of the photodetector is desired. In other words, the circuit as originally designed turns a light on when the cad cell is in darkness and turns the light off when the cell is in the presence of light. The change involves altering the circuit so that the light will turn on when the cad cell is in the presence of light and off when it is in darkness, figure 44-8.

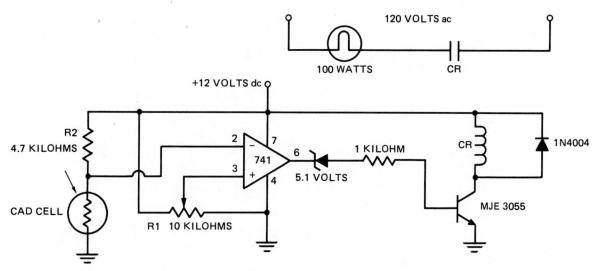

Figure 44-8 Relay turns on when the cad cell is in light.

Examine this circuit and notice that it is basically the same as the circuit shown in figure 44-7. The only differences are that:

1. resistor R1 has been connected to the noninverting input, and
2. the voltage divider, formed by the 4.7-kilohm resistor and the cad cell, has been connected to the inverting input.

Now assume that resistor R1 has been adjusted to produce 5 volts at the noninverting input. When the cad cell is in the presence of light, its resistance is low. A voltage of less than 5 volts is applied to the inverting input. Therefore, the output of the op amp is high and CR relay is turned on. When the cad cell is in darkness, its resistance increases above 4.7 kilohms. This causes a voltage greater than 5 volts to be applied to the inverting input. The output of the op amp goes to a low state and turns off the CR relay. Notice that this circuit causes the light to turn on when the cad cell is in the presence of light and off when it is in darkness. With a few simple changes to the original circuit, there is now a circuit with the opposite sense of operation.

UNIT 45
Design of Circuit #2

CIRCUIT REQUIREMENTS

When a circuit which is large and complex is to be designed, it is generally designed in blocks. This next circuit has been chosen to illustrate this principle. Originally, this circuit was designed as an automatic animal feeder, but with a few changes, it could be used for many applications. The requirements for it are as follows:

1. It must be operated by battery so it can be used in remote locations.
2. It must operate twice each day.
3. Each time it activates, it will run for three seconds and turn off.

CIRCUIT EXPLANATION

A short explanation of the circuit will aid in the understanding of the requirements.

- This circuit is to be used to operate a small dc motor.
- The motor is connected to a metal disk.
- The disk has flat metal blades attached to it.
- Grain is placed into a hopper which is located above the disk.
- When the motor runs, the disk slings grain in all directions to feed the animals.

MEETING THE REQUIREMENTS

Requirement #1 can be met by using components that require only a small amount of power to operate. This circuit was designed to operate for at least 30 days on a 12-volt automobile battery with a capacity of about 60 amp hours.

Requirement #2 is more difficult. When thinking of an operating device that functions twice each day, the time clock usually comes to mind. More specifically, a digital alarm clock is the first thought to meet this requirement. Unfortunately, a common digital alarm clock will not meet it. This circuit requires a device which will produce one pulse each time it alarms. This requirement can be met, then, by using the sun as a clock, and designing the circuit so that the motor will operate in the morning and again

OBJECTIVES

After studying this unit the student should be able to:
- Discuss designing circuits in blocks and then putting the blocks together to form the required circuit
- Construct a complex circuit by building the circuit in blocks and then connecting the blocks together

215

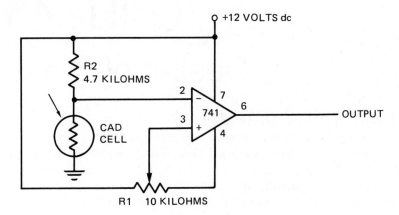

Figure 45-1 Photodetector circuit

in the evening. Since the sun is to be used as the timer, some type of photodetector will be required to sense when the sun rises and sets. The photodetector that was designed in unit 44 should be able to perform this job.

Resistor R1 in figure 45-1 is used to preset a voltage of some value between +12 and 0 volt at the noninverting input. It adjusts the light sensitivity of the photodetector. The 4.7-kilohm resistor and the cad cell form a voltage divider which is connected to the inverting input. When the cad cell is in the presence of light, its resistance is low. This permits a low voltage to be present at the inverting input. This voltage should be less than the preset voltage applied to the noninverting input. Since the voltage applied to the noninverting input is at a higher level than that applied to the inverting input, the output voltage goes high.

In the presence of darkness the resistance of the cad cell increases in value. This causes the voltage applied to the inverting input to increase also. When the voltage applied to the inverting input becomes higher than the voltage applied to the noninverting input, the output changes to a low state. Notice that when the cad cell is in the presence of light the output is high and when in darkness the output is low.

Requirement #3 states that the motor should operate for only three seconds each time it is activated. If this requirement is to be met a timer must be used.

In the circuit of figure 45-2, the photodetector has been coupled to a 555 timer. Recall that in the operation of the 555 timer, the trigger, pin #2, must be connected to a voltage less than one-third of Vcc to trigger the unit. When the timer is triggered, the output, pin #3, turns on. The discharge, pin #7, turns off. When the output turns on, transistor Q1 turns on and energizes the CR

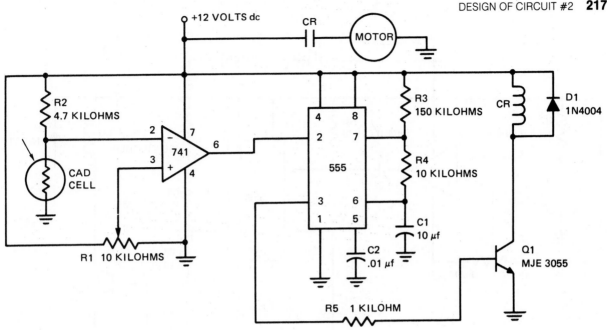

Figure 45-2 Photodetector connected directly to a 555 timer circuit

relay. When the discharge turns off, the capacitor connected to the threshold, pin #6, begins to charge through resistors R3 and R4. When it has been charged to a voltage greater than two-thirds of Vcc, the output attempts to turn off and the discharge to turn on. This cannot happen, however, because the trigger is now connected to a voltage less than one-third of Vcc. The timer remains permanently triggered as long as the output of the op amp is low. The trigger must be *pulsed* if the timer is to operate as a monostable device.

If this part of the circuit is to operate, the trigger must receive a low pulse when the op amp changes from a high to a low state. Adding the .1 μf capacitor and the 100-kilohm resistor, figure 45-3, permits the trigger to receive a one-time pulse when the output of the op amp changes from a high to a low state. The 100-kilohm resistor, R6, charges capacitor C3 to 12 volts. This 12-volt charge keeps the trigger of the 555 timer connected to Vcc. When the output of the op amp changes from a high to a low state, capacitor C3 is suddenly discharged. Resistor R6 permits the capacitor to recharge and again supply 12 volts to the trigger of the timer.

If an oscilloscope is connected to the trigger pin of the timer, it indicates a value of 12 volts until the op amp changes to a low state. When the change takes place, the voltage will pulse low and then return to 12 volts. A waveform similar to the one in figure 45-4 can be seen on the display of the oscilloscope.

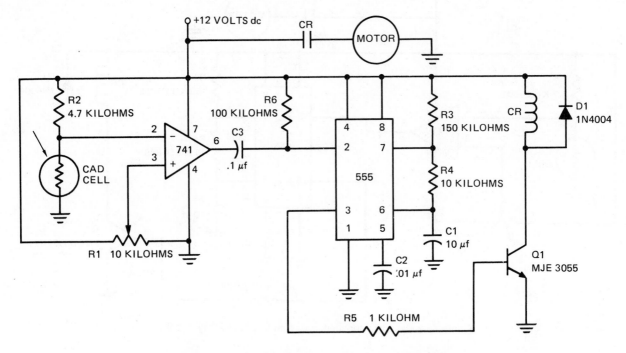

Figure 45-3 R6 and C3 has been added to produce a pulse

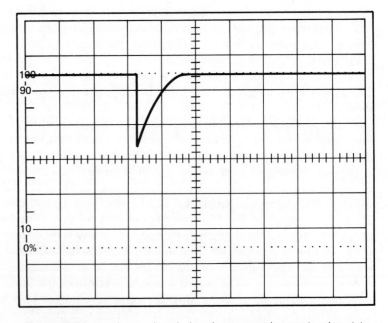

Figure 45-4 Low pulse produced when the op amp changes to a low state

COPYRIGHT © 1983, TEKTRONIX, INC. REPRODUCED BY PERMISSION

This addition to the circuit permits the op amp to provide a single low pulse to the trigger of the timer each time the op amp changes from a high to a low state. When the trigger of the timer receives a low pulse from the op amp, the timer output turns on and energizes the CR relay. When the capacitor that is connected to the threshold charges to two-thirds of Vcc, the output turns off and the discharge turns on. The timer remains in this state until the trigger receives another low pulse from the op amp.

This circuit will now operate the motor for approximately three seconds each time the op amp changes from a high to a low state or when daylight changes to dark. A requirement of the circuit, however, states that the timer must operate twice each day. A method must be found to trigger a timer when the photodetector changes from darkness to light.

When the op amp changes from darkness to light, the output changes from a low to a high state. The 555 timer, however, will

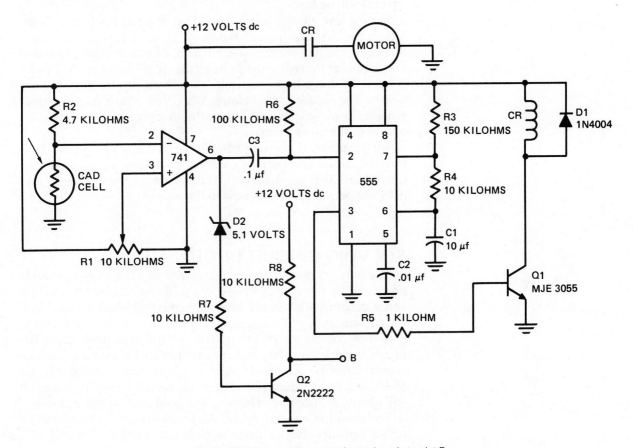

Figure 45-5 An opposite output is produced at point B.

not operate when triggered by a high pulse. Therefore, some method must be found to convert the high pulse of the op amp into a low pulse.

The addition of the components shown in figure 45-5 should produce a low pulse at point B when the output of the op amp changes from a low to a high state. When the output of the op amp is low, the voltage is less than the reverse breakdown voltage of the zener diode. No current can flow to the base of transistor Q2. Since Q2 is turned off, the output voltage at point B is 12 volts. When the output of the op amp changes from a low to a high state, the voltage becomes greater than the reverse breakdown voltage of the zener diode and current flows to the base of transistor Q2. When Q2 turns on, the voltage at point B drops to about .7 volt. Notice that the voltage at point B is opposite the output of the op amp. When the output of the op amp is low, the voltage at point B is high. When the output of the op amp is high, the voltage at point B will be low.

Since a low voltage is now produced at point B when the photodetector changes from darkness to light, it becomes a simple matter to connect another timer circuit. Notice in figure 45-6 that the circuit for timer B is the same as the circuit for timer A. Diodes D3 and D4 have been connected to the output terminals of the timers. The diodes form a simple OR circuit which permits either timer to turn on transistor Q1 without causing feedback to the output of the other timer.

For example, if the output of timer A turns on, diode D3 is forward biased and permits current flow to the base of transistor Q1. When D3 is conducting, D4, however, is reverse biased and does not permit current to flow to the output of timer B.

SOLVING CIRCUIT PROBLEMS

The only problem that may occur with this circuit will probably be caused by the motor. Since it is a dc motor, it contains brushes and a commutator. As the brushes make and break contact with the commutator, voltage spikes are produced. These can be interpreted as commands by the timers and will sometimes cause the timers to retrigger. This, in turn, could cause the motor to operate several times in succession before it stops running. Notice in figure 45-6 that a 50 μf capacitor has been connected across the motor. It will greatly reduce any voltage spikes produced by the motor.

The circuit is now complete. All requirements for its operation have been met. Notice that the circuit is composed mainly of a photodetector and two timers.

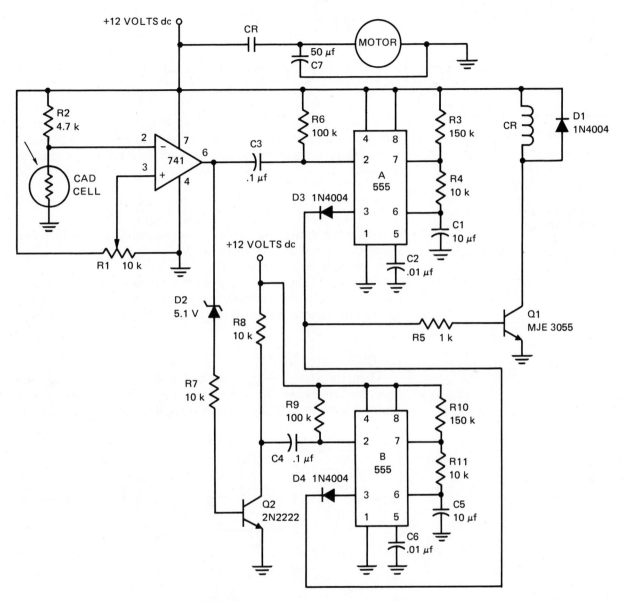

Figure 45-6 Output turns on and off as light changes.

The only real design problem was how to connect the photo-detector circuit to the two timer circuits. As circuits become larger and more complex, it will be found that they are designed by first constructing the major portions of the circuit as blocks and then connecting the blocks together.

UNIT 46
Design of Circuit #3

OBJECTIVES

After studying this unit the student should be able to:

• Discuss the operation of a stop-start pushbutton control
• Discuss the use of logic gates to construct a stop-start pushbutton circuit
• Construct a stop-start pushbutton circuit using logic gates

In this unit a digital circuit will be designed to perform the same function as a common relay circuit. The relay circuit to be used is a basic stop-start pushbutton circuit with overload protection, figure 46-1.

THE COMMON RELAY CIRCUIT

Before beginning the design of an electronic circuit that performs the same function as the relay circuit, the operation of the relay circuit should be discussed.

In the circuit shown in figure 46-1, no current can flow to relay coil M because of the normally open start button and the normally open contact controlled by M relay coil.

Circuit Operation

When the start button is pushed, current flows through the relay coil and normally closed overload contact to the power source, figure 46-2. When current flows through relay coil M, the

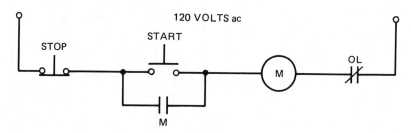

Figure 46-1 Start-stop pushbutton circuit

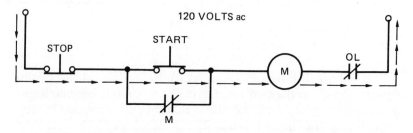

Figure 46-2 Start button energizes M relay coil

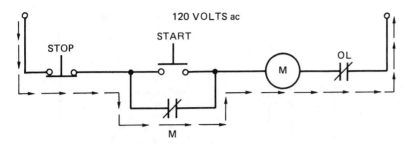

Figure 46-3 M contacts maintain the circuit

contacts connected in parallel with the start button close. These contacts maintain the circuit to coil M when the start button is released and returns to its open position, figure 46-3.

The circuit continues to operate until the stop button is pushed and breaks the circuit to the coil, figure 46-4. When the current flow to the coil stops, the relay deenergizes and contact M reopens. Since the start button and contact M are open, there is no complete circuit to the relay coil when the stop button is returned to its normally closed position. To restart the relay, the start button must be pushed again to provide a complete circuit to the coil of the relay.

The only other logic condition that can occur in this circuit will be caused by the motor connected to the load contacts of M relay. The motor is connected in series with the heater of an overload relay, figure 46-5. When M coil energizes, it closes the load contact M shown in figure 46-5. When this contact closes, it connects the motor to the 120-volt ac power line. If the motor becomes overloaded, it causes too much current to flow through the circuit. When a current greater than normal flows through the overload heater, it causes the heater to produce more heat than it does under normal conditions. If the current becomes high enough, it

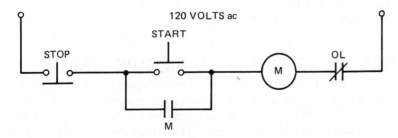

Figure 46-4 Stop button breaks the circuit

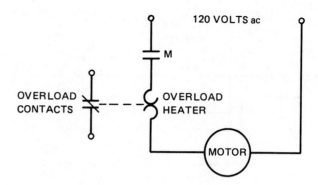

Figure 46-5 Overload relay

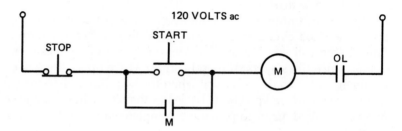

Figure 46-6 Overload contacts break the circuit

causes the normally closed overload contact to open. Notice that this contact is electrically isolated from the heater. It can, therefore, be connected to a totally different voltage source than the motor.

If the overload contact should open, the control circuit is broken and the relay deenergizes, figure 46-6. The open contact affects the circuit in the same way as pushing the stop button. After it has been reset to its normally closed position, the coil remains deenergized until the start button is again pressed.

DESIGNING THE DIGITAL-LOGIC CIRCUIT

Now that the logic of the circuit is understood, the design of a digital-logic circuit that will operate in this manner can begin.

The first problem is to find a circuit that will turn on with one pushbutton and off with another. The circuit shown in figure 46-7 can do this.

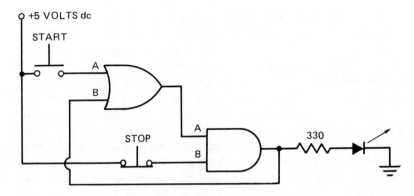

Figure 46-7 Output turns on and off with pushbuttons

Operation of the Circuit

This circuit consists of an OR and an AND gate, and functions as follows:

1. Input A of the OR gate is connected to a normally open pushbutton, which is connected to +5 volts dc.
2. The B input of the OR gate is connected to the output of the AND gate.
3. The output of the OR gate is connected to the A input of the AND gate.
4. The B input of the AND gate is connected through a normally closed pushbutton to +5 volts dc.
5. This pushbutton is used as the stop button.
6. The output of the AND gate is the output of the circuit.

Understanding the Logic of the Circuit

To understand the logic of this circuit, assume that the output of the AND gate is low. This produces a low at input B of the OR gate. Since the pushbutton connected to the A input is open, a low is produced at this input also. (Note: When all inputs of an OR gate are low, its output is low also.) The low output of the OR gate is connected to the A input of the AND gate. The B input of the AND gate is connected to a high through the normally closed pushbutton switch. Since the A input of the AND gate is low, its output is forced to remain in a low state.

When the start button is pushed, a high is connected to the A input of the OR gate. This causes the output of the OR gate to change to high. The high is connected to the A input of the AND gate. This gate now has both its inputs high, so its output changes from a low to a high state. When the output of the AND gate changes to a high state, the B input of the OR gate becomes high also. Since the OR gate now has a high connected to its B input, its output will remain high when the pushbutton is returned to its open condition and input A becomes low. Notice that this circuit operates the same as the relay circuit when the start button is pushed. The output changes from a low to a high state and the circuit locks in this condition so the start button can be reopened.

When the normally closed stop button is pushed, the B input of the AND gate changes from a high to a low. When it changes to a low state, the output of the AND gate changes to a low state also. This causes a low to appear at the B input of the OR gate. This gate now has both inputs low, so its output changes from a high to a low state. Since the A input of the AND gate is now low, the output is forced to remain low when the stop button returns to its closed position and a high is connected to the B input. A circuit now exists that can be turned on with the start button and off with the stop button.

Adding the Overload Contact

The next design problem is to connect the overload contact into the circuit. It must be done in such a way that it causes the output of the circuit to turn off if the overload contact should open. The first impulse probably is to connect the circuit as shown in figure 46-8.

In this circuit, the output of AND gate #1 is connected to the A input of AND gate #2. The B input of AND gate #2 is connected to high through the normally closed overload contact. If the overload contact remains closed, the B input remains high. The output of AND gate #2 is, therefore, controlled by the A input. If the output of AND gate #1 changes to a high state, the output of AND gate #2 changes to a high state also. If the output of AND gate #1 becomes low, the output of AND gate #2 will become low too.

If the output of AND gate #2 is high and the overload contact opens, the B input becomes low and the output changes from a high to a low state. This circuit appears to operate with the same logic as the relay circuit until the logic is closely examined.

Assume that the overload contact is closed and the output of AND gate #1 is high. Since both inputs of AND gate #2 are high, the output is high also. Now assume that the overload contact opens and causes B input to change to a low condition. This forces the output of AND gate #2 to change to a low state also. The A input of AND gate #2 is still high, however. If the overload contact is reset, the output immediately changes back to a high state. If

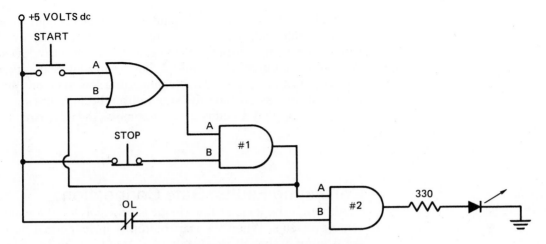

Figure 46-8 Overload can turn off output

the overload contact opens and is then reset in the relay circuit, the relay will not restart itself. The start button has to be pushed to restart the circuit. Although this is a small difference in circuit logic, it could become a safety hazard in some cases.

CORRECTING FAULTS: DESIGN CHANGE

This fault can be corrected with a slight design change, figure 46-9. In this circuit, the normally closed stop button is connected to the A input of AND gate #2, and the normally closed overload switch is connected to the B input. As long as both of these inputs

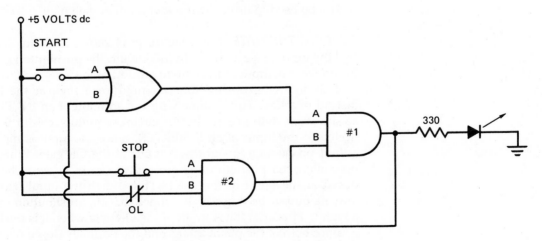

Figure 46-9 Correct circuit operation

are high, the output of AND gate #2 provides a high to the B input of AND gate #1. If either the stop button or the overload contact opens, the output of AND gate #2 changes to a low state. When the B input of AND gate #2 changes to a low state, it causes the output of AND gate #1 to change to a low state and unlock the circuit just as pushing the stop button did in figure 46-8. The logic of this digital circuit is now the same as the relay circuit.

Correcting Faults: Gate Connections

Although the logic of this circuit is now correct, there are still some problems. When the start button is in its normal position, the A input of the OR gate is not connected to anything. When an input is left in this condition, the gate may not be able to determine if the input should be high or low. The gate could, therefore, assume either condition. To prevent this, gate inputs must always be connected to a definite high or low.

When using TTL logic, inputs are always pulled high with a resistor as opposed to being pulled low. If a resistor is used to pull an input low as shown in figure 46-10, it causes the gate to have a voltage drop at its output. This means that in the high state, the output of the gate may be only 3 or 4 volts instead of 5 volts. If this output is used as the input of another gate, and this other gate has been pulled low with a resistor, the output of this second gate may be only 2 or 3 volts. Notice that each time a gate is pulled low through a resistor, its output voltage becomes low. If this is done through several steps, the output voltage soon becomes so low it cannot be used to drive the input of another gate high.

PULLING THE GATE HIGH: Figure 46-11 shows a resistor used to pull the input of a gate high. In this circuit, the pushbutton is used to connect the input of the gate to ground or low.

If desired, the pushbutton can produce a high at the input instead of a low. This is done with the addition of an INVERTER as shown in figure 46-12. In this circuit, a pullup resistor is connected to the input of an INVERTER. Since the input is high, its output produces a low at the A input of the OR gate. When the normally open pushbutton is pressed, it causes a low to be produced at the input of the INVERTER. When the input becomes low, its output becomes high. Notice that the pushbutton causes a high to appear at the A input of the OR gate when it is pushed.

Since both of the pushbuttons and the normally closed overload contact are used to provide high inputs, the circuit will be changed

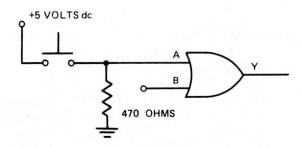

Figure 46-10 Input pulled low with a resistor

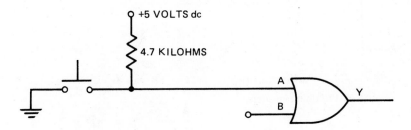

Figure 46-11 Resistor used to pull the input of a gate high

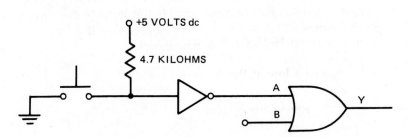

Figure 46-12 Pushbutton produces a high at the input

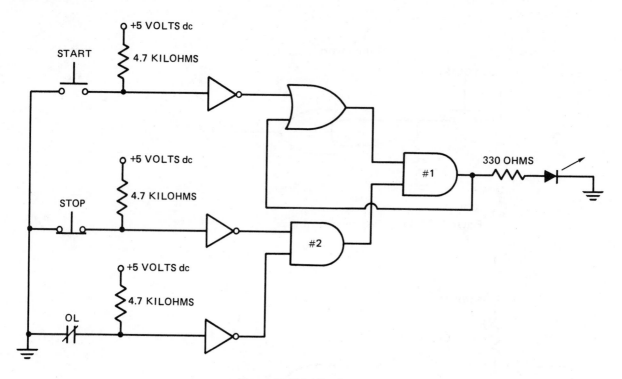

Figure 46-13 Adding inverters

as shown in figure 46-13. Notice that the normally closed pushbutton and overload switch connected to the inputs of AND gate #2 are connected to ground instead of Vcc. When they are connected to ground, a low is provided to the input of the INVERTERS they are connected to. The INVERTERS, therefore, produce a high at the input of the AND gate. If one of these normally closed switches is opened, a high is provided at the input of the INVERTER gate. This causes its output to become low. If the logic of the circuit shown in figure 46-13 is checked, it can be seen that it is the same as the logic of the circuit shown in figure 46-9.

Correcting Faults: The Output

The final design problem for this circuit concerns the output. So far, an LED has been used as the load. It is used to indicate when the output is high and low. The original circuit, however, was used to control a 120-volt ac motor.

Circuit
Applications

UNIT 47

The Design of Circuit #4

OBJECTIVES

After studying this unit the
student should be able to:

- Discuss the operation of an
electronic lock
- Construct an electronic lock
using digital-logic gates

CIRCUIT REQUIREMENTS

In this circuit, logic gates will be used to design an electronic lock. The requirements are:

1. The opening of the lock occurs by pressing four numbers in the proper sequence.
2. All pushbuttons are normally open and one side of each button is connected to ground. This permits the use of a 10-digit keyboard with a common ground connection.
3. If any number which is not used in the combination is pressed, it will reset the circuit, making it necessary to begin again with the first number of the sequence.

For this circuit, the numbers 2, 4, 6, and 8 will be used as the combination.

NAND GATES USED AS A LATCH

The first part of this circuit can be constructed with two NAND gates used as a latch, figure 47-1.

Understanding the Circuit

To understand the logic of this circuit, assume the output of NAND gate #1 to be low. This provides a low to the A input of NAND gate #2. Since a NAND gate has a high output if any input is low, the output of NAND gate #2 provides a high for the B input of NAND gate #1. The A input of NAND gate #1 is forced high by the 4.7-kilohm resistor connected to +5-volts dc. The output of NAND gate #1 remains low and the output of NAND gate #2 remains high. If pushbutton #2 is pressed, the A input of NAND gate #1 is driven low. This forces the output of NAND gate #1 to change to a high state. When the A input of NAND gate #2 changes to high, its output becomes low and provides a low to input B of NAND gate #1. When the B input of NAND gate #1 is forced low, the output of NAND gate #1 remains in a high

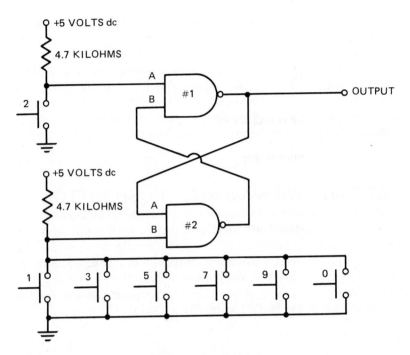

Figure 47-1 Latch circuit

state when pushbutton #2 returns to its open position. When the output of NAND gate #1 changes to a high state, it forces the A input of NAND gate #2 to become high also. NAND gate #2 now has both of its inputs high. This forces the output to remain low.

If pushbutton #1, 3, 5, 7, 9, or 0 is pushed, the B input of NAND gate #2 is connected to ground or low. When the B input of NAND gate #2 becomes low, its output changes to a high state. This provides a high to the B input of NAND gate #1. Both inputs of NAND gate #1 are now high. This forces the output to change to a low state. When the output of NAND gate #1 becomes low, it provides a low to the A input of NAND gate #2. Since the A input of NAND gate #2 is now low, it forces the output to remain high when pushbutton #1, 3, 5, 7, 9, or 0 is released and returns to its open position.

Fulfilling Requirement #3
If the output of NAND gate #1 is used as the circuit output, a circuit exists that has a high output when pushbutton #2 is

pressed, and a low output when pushbutton #1, 3, 5, 7, 9, or 0 is pressed. This fulfills the part of the circuit requirement which states that if any pushbutton not used in the combination is pressed, it will reset the circuit. It is necessary to begin again with the first number of the combination sequence. The circuit in figure 47-1 shows the connection of the six pushbuttons not used in the combination sequence. For the remainder of this unit, one pushbutton, labeled N, will be used to represent these six pushbuttons.

LATCHING WITH A SINGLE PUSHBUTTON

The next problem is to design a circuit that can be latched with a single pushbutton only after it receives a high input from the output of NAND gate #1. This would normally be considered a simple circuit to design, but the requirement of using pushbuttons with common grounds makes it more difficult than it first appears. There are probably several ways this can be accomplished, but the circuit shown in figure 47-2 is the circuit that will be used.

Circuit Logic

To understand the logic of this circuit, assume the output of the AND gate to be low. If the output of the AND gate is low, the B input of the OR gate is low also. The A input of the OR gate is connected to the output of an INVERTER. The input of the IN-

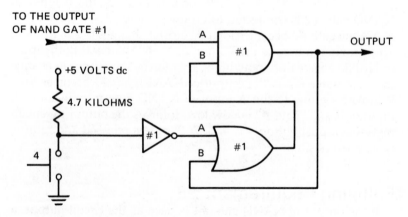

Figure 47-2 Latching with a single pushbutton

VERTER is connected to high through a 4.7-kilohm resistor connected in turn to +5-V dc. Since the input of the INVERTER is high, its output is low. With both inputs of the OR gate low, its output provides a low to the B input of the AND gate. Since this input is low, the output is forced to remain low even if the A input is provided with a high from the output of NAND gate #1. Now assume that the A input of the AND gate has been driven high by the output of NAND gate #1. If pushbutton #4 is pressed, the input of the INVERTER is connected to ground or low. This forces the output of the INVERTER to change and provides a high to the A input of the OR gate. When this input becomes high, the output of the OR gate changes and provides a high to the B input of the AND gate. Since both inputs of the AND gate are now high, its output changes from a low to a high. When this output becomes high, it provides a high to the B input of the OR gate. The high connected to the B input forces the output of the OR gate to remain high when pushbutton #4 is released and the A input becomes low again. Once the AND gate has been turned on, its output remains in the high state until the output of NAND gate #1 changes state and forces the A input of the AND gate to change to a low state.

Creating the Combination

An electronic lock now exists with a two-number combination. If an output is to be obtained at the AND gate, pushbutton #2 must first be pressed, followed by pushbutton #4. Using the circuit shown in figure 47-2 as a building block, it becomes a simple matter to add the other two numbers of the combination. When the output of the AND gate in figure 47-2 is used as the input of another identical circuit, the rest of the circuit becomes a simple chain. In the circuit of figure 47-3, the input of a block is provided by the output of the block preceding it. For instance, AND gate #3 can not have a high output unless AND gate #2 provides a high for its A input. AND gate #2 cannot have a high output unless AND gate #1 provides a high to its A input. Notice that each of the pushbuttons used in the combination must be pressed in the proper sequence before AND gate #3 provides a high output.

This circuit uses an LED to indicate a high or low output. In actual practice, however, the output of AND gate #3 could be connected to the base of a transistor. The transistor could be used to control the coil of a relay, or the output of AND gate #3 could be connected to the input of a solid-state relay. The solid-state relay could then be used to control almost anything.

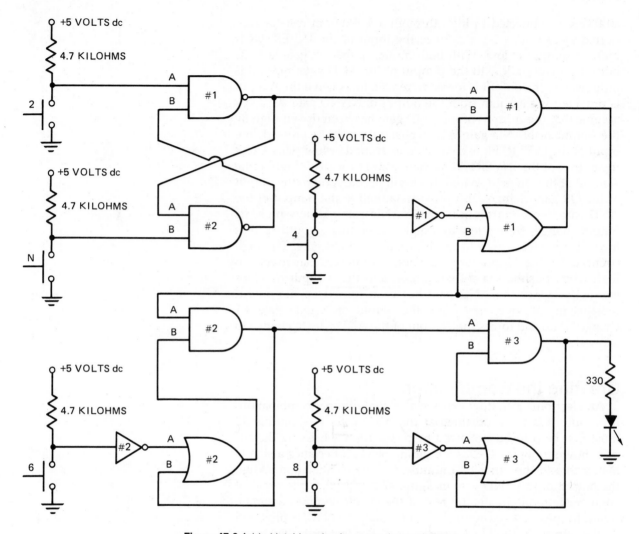

Figure 47-3 Added latching circuits create the combination

CIRCUIT CONNECTION

Until now the logic circuits that have been connected were relatively simple, using only two or three gates. If the INVERTERS are counted as gates, this circuit requires a connection of eleven different gates. Although only 4 ICs are required to construct the circuit, it is easy to become confused when making so many connections in a small space. It is helpful to number the schematic with the pin numbers of the IC. If a connection diagram of the IC is used as shown in unit 42, it is a simple matter to number the gate connections with the proper pin numbers of the IC, figure 47-4.

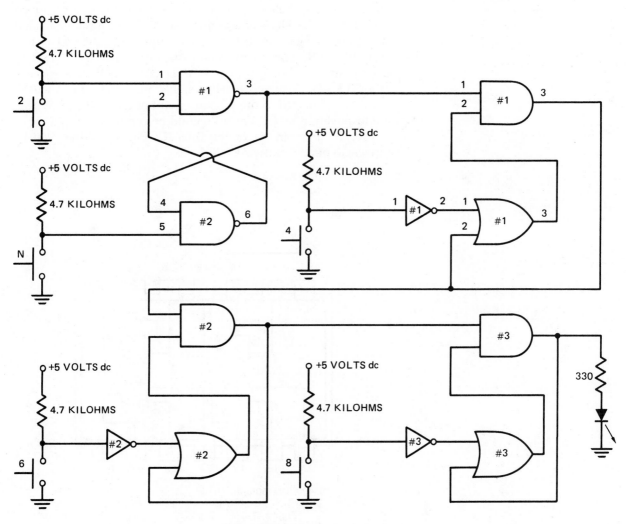

Figure 47-4 Adding grate pin numbers

Numbering the Gates

In figure 47-5, the connection diagram for a 7400N IC is shown. The 7400N is a quad two-input NAND gate. Notice in figure 47-5 that NAND gate #1 has been labeled with a #1 and #2 at its inputs and a #3 at its output. The connection diagram in figure 47-5 shows that one of the NAND gates in the 7400 IC has its inputs connected to pins #1 and #2 and its output is connected to pin #3. NAND gate #2 in figure 47-4 has its inputs labeled #4 and #5 and its output is labeled #6. The connection diagram in figure 47-5 shows the second gate in the 7400 IC has its inputs connected to pins #4 and #5. Its output is connected to pin #6.

If each gate is labeled in this manner, connection becomes much easier. Figure 47-6 shows the connection diagrams for a 7408, quad two-input AND gate; a 7432, quad two-input OR gate; and a 7404, hex INVERTER. These connection diagrams have been used to label the input and output pins of AND gate #1, OR gate #1, and INVERTER #1. Figure 47-4 shows that pin #1 of the 7400 IC should be connected to the junction point of the 4.7-kilohm resistor and pushbutton #2. Pin #3 should be connected to pin #4 of the 7400 IC and pin #1 of the 7408 IC. Pin #6 of the 7400 IC connects to pin #2 of the 7400 IC.

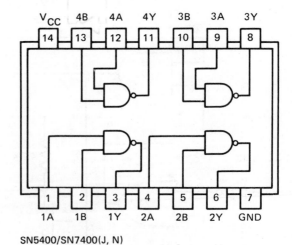

SN5400/SN7400(J, N)

Figure 47-5 NAND gate IC

Notice that connection becomes much simpler if the schematic is first labeled with the proper IC pin numbers. They are then used as a guide when connecting the circuit.

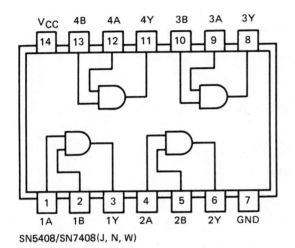

SN5408/SN7408(J, N, W)

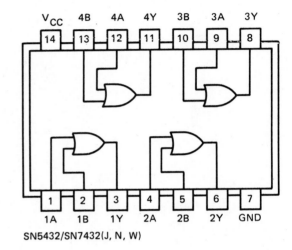

SN5432/SN7432(J, N, W)

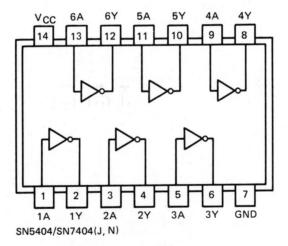

SN5404/SN7404(J, N)

Figure 47-6 IC diagrams

COURTESY OF TEXAS INSTRUMENTS, INC. © 1973

APPENDIX A
Testing Solid-state Components

PROCEDURE 1: TESTING A DIODE

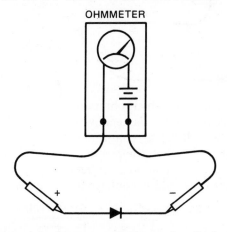

Step 1. Connect the ohmmeter leads to the diode. Notice if the meter indicates continuity through the diode or not.

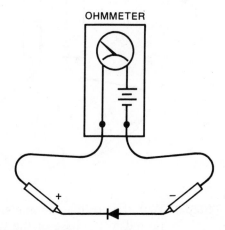

Step 2. Reverse the diode connection to the ohmmeter. Notice if the meter indicates continuity through the diode or not. The ohmmeter should indicate continuity through the diode in only one direction. (Note: If continuity is not indicated in either direction, the diode is open. If continuity is indicated in both directions, the diode is shorted.)

PROCEDURE 2: TESTING A TRANSISTOR

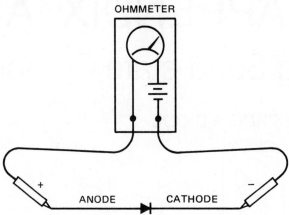

Step 1. Using a diode, determine which ohmmeter lead is positive and which is negative. The ohmmeter will indicate continuity through the diode only when the positive lead is connected to the anode of the diode and the negative lead is connected to the cathode.

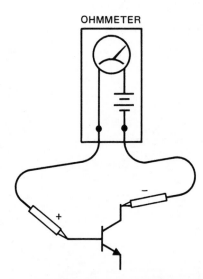

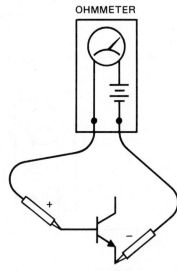

Step 2. If the transistor is an NPN, connect the positive ohmmeter lead to the base and the negative lead to the collector. The ohmmeter should indicate continuity. The reading should be about the same as the reading obtained when the diode was tested.

Step 3. With the positive ohmmeter lead still connected to the base of the transistor, connect the negative lead to the emitter. The ohmmeter should again indicate a forward diode junction. (Note: If the ohmmeter does not indicate continuity between the base-collector or the base-emitter, the transistor is open.)

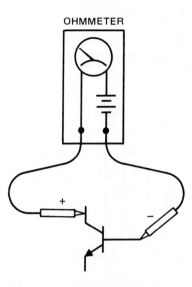

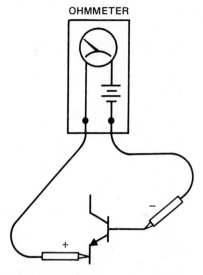

Step 4. Connect the negative ohmmeter lead to the base and the positive lead to the collector. The ohmmeter should indicate infinity or no continuity.

Step 5. With the negative ohmmeter lead connected to the base, reconnect the positive lead to the emitter. There should again be no indication of continuity. (Note: If a very high resistance is indicated by the ohmmeter, the transistor is "leaky" but may still operate in the circuit. If a very low resistance is seen, the transistor is shorted.)

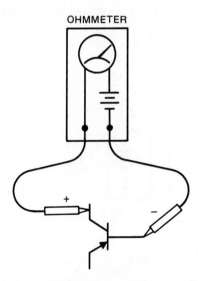

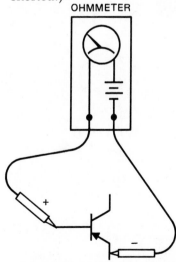

Step 6. To test a PNP transistor, reverse the polarity of the ohmmeter leads and repeat the test. When the negative ohmmeter lead is connected to the base, a forward diode junction should be indicated when the positive lead is connected to the collector or emitter.

Step 7. If the positive ohmmeter lead is connected to the base of a PNP transistor, no continuity should be indicated when the negative lead is connected to the collector or the emitter.

PROCEDURE 3: TESTING A UNIJUNCTION TRANSISTOR

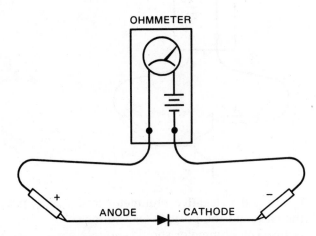

Step 1. Using a junction diode, determine which ohmmeter lead is positive and which is negative. The ohmmeter will indicate continuity when the positive lead is connected to the anode and the negative lead is connected to the cathode.

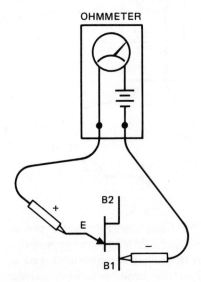

Step 2. Connect the positive ohmmeter lead to the emitter lead and the negative lead to base #1. The ohmmeter should indicate a forward diode junction.

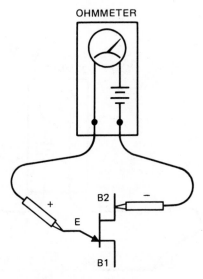

Step 3. With the positive ohmmeter lead connected to the emitter, reconnect the negative lead to base #2. The ohmmeter should again indicate a forward diode junction.

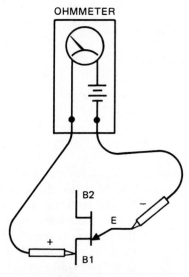

Step 4. If the negative ohmmeter lead is connected to the emitter, no continuity should be indicated when the positive lead is connected to either base #1 or base #2.

PROCEDURE 4: TESTING AN SCR

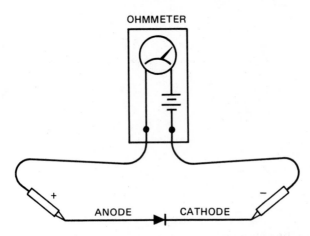

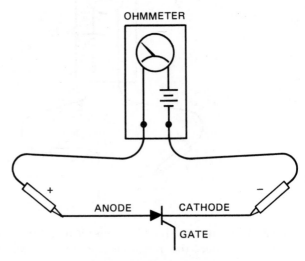

Step 1. Using a junction diode, determine which ohmmeter lead is positive and which is negative. The ohmmeter will indicate continuity only when the positive lead is connected to the anode of the diode and the negative lead is connected to the cathode.

Step 2. Connect the positive ohmmeter lead to the anode of the SCR and the negative lead to the cathode. The ohmmeter should indicate no continuity.

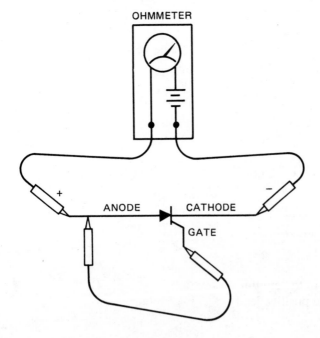

Step 3. Using a jumper lead, connect the gate of the SCR to the anode. The ohmmeter should indicate a forward diode junction when the connection is made. (Note: If the jumper is removed, the SCR may continue to conduct or it may turn off. This will be determined by whether the ohmmeter can supply enough current to keep the SCR above its holding current level or not.)

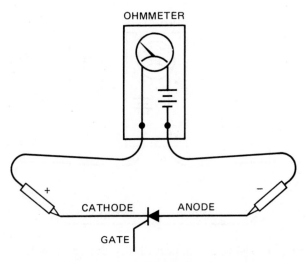

Step 4. Reconnect the SCR so that the cathode is connected to the positive ohmmeter lead and the anode is connected to the negative lead. The ohmmeter should indicate no continuity.

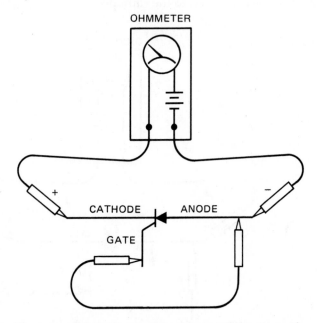

Step 5. If a jumper lead is used to connect the gate to the anode, the ohmmeter should indicate no continuity. (Note: SCRs designed to switch large currents (50 amperes or more) may indicate some leakage current with this test. This is normal for some devices.)

PROCEDURE 5: TESTING A TRIAC

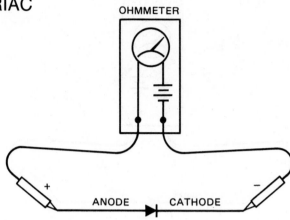

Step 1. Using a junction diode, determine which ohmmeter lead is positive and which is negative. The ohmmeter will indicate continuity only when the positive lead is connected to the anode and the negative lead is connected to the cathode.

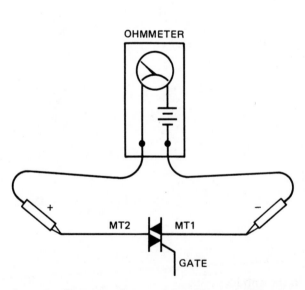

Step 2. Connect the positive ohmmeter lead to MT2 and the negative lead to MT1. The ohmmeter should indicate no continuity through the triac.

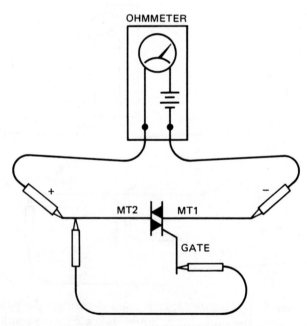

Step 3. Using a jumper lead, connect the gate of the triac to MT2. The ohmmeter should indicate a forward diode junction.

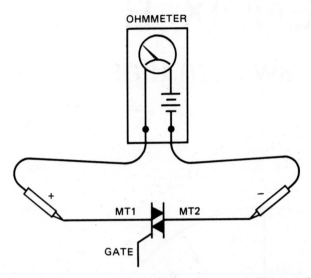

Step 4. Reconnect the triac so that MT1 is connected to the positive ohmmeter lead and MT2 is connected to the negative lead. The ohmmeter should indicate no continuity through the triac.

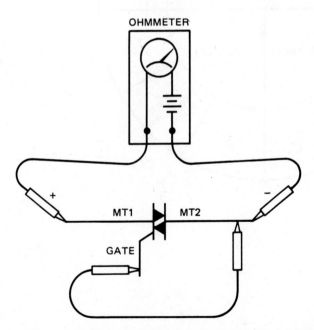

Step 5. Using a jumper lead, again connect the gate to MT2. The ohmmeter should indicate a forward diode junction.

APPENDIX B
Ohm's Law Formulas

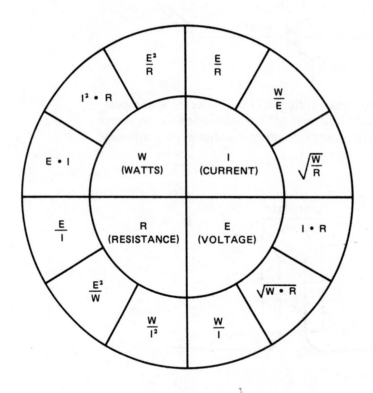

APPENDIX C
Parts List

DIODES:	Light emitting 1N4004 Silicon 400PRV 1 A 12-volt zener 5 watts 5.1-volt zener .25 watt

TRANSISTORS:	2N2222 NPN Silicon .5 watt Vce 30 volts
	   IC 150 ma
	2N2907 PNP Silicon .5 watt Vce 30 volts
	   IC 150 ma
	2N5039 NPN Silicon 140 watts Vce 120
	   IC 10 A
	2N2646 UJT 300 mw
	MJE 3055 NPN Silicon 90 watts Vce 60
	   IC 4 A

FET	2N5458 N-Channel JFET
SCRs:	2N4170 200 volts 8 A Igate 20 ma Egate 1.5 volts 2N1598 300 volts 1.6 A
DIAC:	1N5761 Turn on 32 volts Turn off 25 volts Imax 2 A
TRIAC:	2N6151 200 volts 10 A
SOLID-STATE RELAY:	General Electric #CR120SR110D 240 volts ac 10 A (or) International Rectifier #D2410
INTEGRATED CIRCUITS:	555 Timer 741 Operational amplifier 7400N Quad 2-input NAND gate TTL 7402N Quad 2-input NOR gate TTL 7404N hex INVERTER TTL 7408N Quad 2-input AND gate TTL 7432N Quad 2-input OR gate TTL

TRANSFORMERS:	117/117 volts isolation transformer 2A
	117/24 volts center tapped 3 A
	117/12.6 volts center tapped 3 A
RELAY:	12 volts dc Potter Brumfield or equivalent
RESISTORS (one-half watt):	10, 27, 100, 330, 470, 1 k, 1.2 k, 1.5 k, 2.2 k, 3 k, 3.3 k, 4.7 k, 6.8 k, 10 k, 15 k, 22 k, 27 k, 47 k, 68 k, 75 k, 100 k, 220 k, 470 k, 1 meg
RESISTORS (SPECIAL):	200 ohms 2 watts
	100 ohms 2 watts

RESISTORS (SPECIAL):

100 ohm

3.3 k 2 watts   Light Bulb

25 ohms 25 watts

RESISTORS (VARIABLE):	1 k, 5 k, 10 k, 50 k, 250 k, 1 meg, 2 meg
SWITCHES:	DPDT, SPST, Pushbuttons (NO and NC)
CAPACITORS (50 volt):	.01 µf, .1 µf, 1 µf, 2 µf, 5 µf, 10 µf, 25 µf, 50 µf, 100 µf, 150 µf, 200 µf, 470 µf, 1000 µf
CAD CELL	Radio Shack or equivalent
PIEZO BUZZER	Sonalert #SC628 or equivalent

APPENDIX D
Resistor Color Code

BLACK	0
BROWN	1
RED	2
ORANGE	3
YELLOW	4
GREEN	5
BLUE	6
VIOLET	7
GRAY	8
WHITE	9

TOLERANCE

GOLD	5%
SILVER	10%
NONE	20%

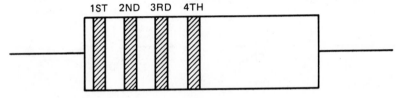

1ST BAND—NUMBER
2ND BAND—NUMBER
3RD BAND—MULTIPLIER (ADD THAT NUMBER OF 0'S TO THE FIRST TWO NUMBERS)
4TH BAND—TOLERANCE

EXAMPLE: A RESISTOR HAS COLORS OF YELLOW, VIOLET, ORANGE, AND GOLD.
FIRST TWO COLORS ARE YELLOW AND VIOLET, WHICH ARE 4 AND 7. THE THIRD BAND IS ORANGE,
WHICH IS 3. THEREFORE, ADD 3 ZEROS TO THE 47. RESISTANCE IS: 47000 OHMS.
4TH BAND IS GOLD WHICH IS 5% TOLERANCE. THE RESISTOR IS 47000 OHMS ±5%.

APPENDIX E
Schematic Symbols

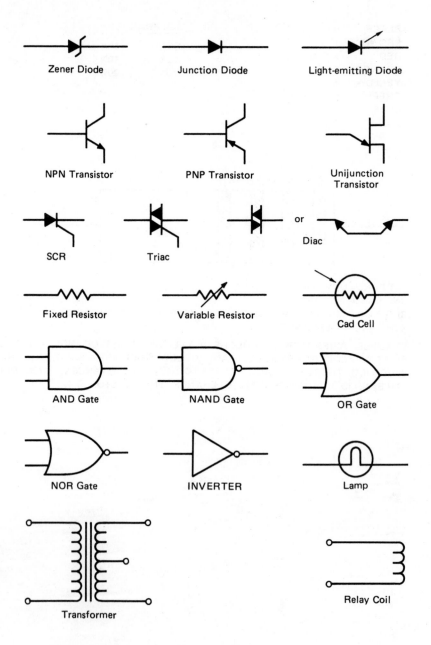

Zener Diode

Junction Diode

Light-emitting Diode

NPN Transistor

PNP Transistor

Unijunction Transistor

SCR

Triac

or

Diac

Fixed Resistor

Variable Resistor

Cad Cell

AND Gate

NAND Gate

OR Gate

NOR Gate

INVERTER

Lamp

Transformer

Relay Coil

APPENDIX F

Conversion Factors for Sine Wave Voltages

TO CHANGE	TO	MULTIPLY BY
PEAK	RMS	.707
PEAK	AVERAGE	.637
PEAK	PEAK TO PEAK	2
RMS	PEAK	1.414
AVERAGE	PEAK	1.567
RMS	AVERAGE	.9
AVERAGE	RMS	1.111

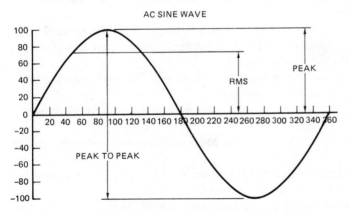

AC SINE WAVE

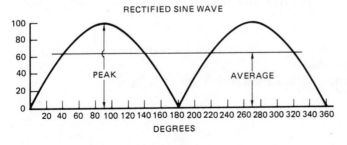

RECTIFIED SINE WAVE

DEGREES

Glossary

AC (alternating current)—Current that reverses its direction of flow periodically. Reversals generally occur at regular intervals.

Alternating current—*See* ac.

Alternator—A machine used to generate alternating current by rotating conductors through a magnetic field.

Amplifier—A device used to increase a signal.

Amplitude—The highest value reached by a signal, voltage, or current.

AND gate—A digital-logic gate that must have all of its inputs high to produce an output.

Anode—The positive terminal of an electronic device.

Applied voltage—The amount of voltage connected to a circuit or device.

Astable mode—The state in which an oscillator can continually turn itself on and off, or continually change from positive to negative output.

Attenuator—A device that decreases the amount of signal, voltage, or current.

Barrier charge—The potential developed across a semiconductor junction.

Base—The semiconductor region between the collector and emitter of a transistor. The base controls the current flow through the collector-emitter circuit.

Base current—The amount of current that flows through the base-emitter section of a transistor.

Bias—A dc voltage applied to the base of a transistor to preset its operating point.

Bounceless switch—A circuit used to eliminate contact bounce in mechanical contacts.

Bridge circuit—A circuit that consists of four sections connected in series to form a closed loop.

Bridge rectifier—A device constructed with four diodes that converts both positive and negative cycles of ac voltage into dc voltage.

Cad cell—A device that changes its resistance with a change of light intensity.

Capacitance—The electrical size of a capacitor.

Capacitor—A device made with two conductive plates separated by an insulator or dielectric.

Capacitive—Any circuit or device having characteristics similar to those of a capacitor.

Cathode—The negative terminal of a device.

Cathode-ray tube (CRT)—An electron beam tube in which the beam of electrons can be focused to any point on the face of the tube. The electron beam causes the face of the tube to produce light when it is struck by the beam.

Center-tapped transformer—A transformer that has a wire connected to the electrical midpoint of its winding. Generally the secondary is tapped.

Charge time—The amount of time necessary to charge a capacitor.

Choke—An inductor designed to present an impedance to ac current, or to be used as the current filter of a dc power supply.

Circuit—An arrangement of electrical devices to form one or more complete paths for current flow.

Clock timer—A time-delay device that uses an electric clock to measure the delay period.

Collapse (of a magnetic field)—Occurs when a magnetic field suddenly changes from its maximum value to a zero value.

Collector—A semiconductor region of a transistor that must be connected to the same polarity as the base.

Comparator—A device or circuit that compares two like quantities, such as voltage levels.

Conduction level—The point at which an amount of voltage or current will cause a device to conduct.

Conductor—A device or material that permits current to flow through it easilly.

CRT—*See* cathode-ray tube.

Current—The rate of the flow of electrons.

Current flow—The flow of electrons.

Current generator—A circuit designed to deliver a certain amount of current as opposed to a certain amount of voltage.

Current rating—The amount of current flow a device is designed to withstand.

Dashpot timer—A device using a piston moving through a liquid to produce a time delay.

DC (direct current)—Current that does not reverse its direction of flow.

DE-MOSFET (depletion-enhancement metal oxide semiconductor field effect transistor)—This transistor can be operated in either a depletion or enhancement mode.

Delta connection—A circuit formed by connecting three electrical devices in series to form a closed loop. It is used most often in three-phase connections.

Diac—A bidirectional diode.

Digital device—A device that has only two states of operation, on and off.

Digital logic—Circuit elements connected in such a manner as to solve problems using components that have only two states of operation.

Digital voltmeter—A voltmeter that uses a direct reading numerical display as opposed to a meter movement.

Diode—A two-element device that permits current to flow through it in only one direction.

Direct current—*See* dc.

E-MOSFET (enhancement metal oxide semiconductor field effect transistor)—This transistor can be operated in the enhancement mode only.

Emitter—The semiconductor region of a transistor that must be connected to a polarity different than the base.

EXCLUSIVE-OR gate—A digital-logic gate that will produce an output when its inputs have opposite states of logic level.

FET–*See* field effect transistor.

Field effect transistor—A transistor that controls the flow of current through it with an electric field.

Filter—A device used to remove the ripple produced by a rectifier.

Frequency—The number of complete cycles of ac voltage that occur in one second.

Gain—The increase in signal power produced by an amplifier.

Gate—A device that has multiple inputs and a single output, or one terminal of some solid-state devices such as SCRs or triacs.

Heat sink—A metallic device designed to increase the surface area of an electronic component for the purpose of removing heat at a faster rate.

Holding contacts—Contacts used for the purpose of maintaining current flow to the coil of a relay.

Holding current—The amount of current needed to keep an SCR or triac turned on.

Hysteresis loop—A graphic curve that shows the value of magnetizing force for a particular type of material.

IGFET (insulated gate field effect transistor)—This device is the same as a MOSFET.

Impedance—The total opposition to current flow in an electrical circuit.

Induced current—Current produced in a conductor by the cutting action of a magnetic field.

Inductor—A coil.

Input voltage—The amount of voltage connected to a device or circuit.

Insulator—A material used to electrically isolate two conductive surfaces.

Internal relay—Digital-logic circuits in a programmable controller that can be programmed to operate in the same manner as control relays.

INVERTER (gate)—A digital-logic gate that has an output opposite its input.

Inverting input—The input terminal of an operational amplifier that will cause the output to assume the opposite polarity of voltage.

Isolation transformer—A transformer whose secondary winding is electrically isolated from its primary winding.

JFET (junction field effect transistor)—A field effect transistor formed by combining layers of semiconductor material. The JFET has an input impedance equal to the reverse bias impedance of a p-n junction. This impedance is approximately 20,000 Meg. ohms.

Junction diode—A diode that is made by joining together two pieces of semiconductor material.

Kick-back diode—A diode used to eliminate the voltage spike induced in a coil by the collapse of a magnetic field.

Lattice structure—An orderly arrangement of atoms in a crystalline material.

LED (light-emitting diode)—A diode that will produce light when current flows through it.

Light-emitting diode—*See* LED.

Magnetic field—The space in which a magnetic force exists.

Microprocessor—A small computer. The central processing unit is generally made from a single integrated circuit.

Mode—A state of condition.

Monostable mode—The state in which an oscillator or timer will operate through only one sequence of events.

MOSFET (metal oxide semiconductor field effect transistor)—A field effect transistor that has no electrical connection between the gate and channel. The input impedance is approximately two billion ohms.

Motor controller—A device used to control the operation of a motor.

N-channel—A field effect transistor so constructed that n-type semiconductor material is used for the channel through the device.

NAND gate—A digital-logic gate that will produce a high output only when all of its inputs are in a low state.

Negative resistance—The property of a device in which an increase of current flow causes an increase of conductance. The increase of conductance causes a decrease in the voltage drop across the device.

Noninverting input—The input of an operational amplifier that produces the same polarity of voltage at the output.

NOR gate—A digital-logic gate that will produce a high output when any of its inputs are low.

Normally closed—The contact of a relay that is closed when the relay coil is deenergized.

Normally open—The contact of a relay that is open when the relay coil is deenergized.

Off-delay timer—A timer whose contacts change position immediately when the coil or circuit is energized, but delay returning to their normal position when the coil or circuit is deenergized.

Ohmmeter—A meter used to measure resistance.

On-delay timer—A timer whose contacts will delay changing position when the coil or circuit is energized, but change back immediately when the coil or circuit is deenergized.

Operational amplifier—A direct-coupled integrated circuit amplifier used for high output current applications.

Op amp—An operational amplifier.

Optoisolator—A device used to connect sections of a circuit by means of a light beam.

Oscillator—A device or circuit that is used to change dc voltage into ac voltage.

Oscilloscope—An instrument that measures the amplitude of voltage with respect to time.

Out-of-phase voltage—A voltage that is not in phase when compared to some other voltage or current.

Output pulse—A short duration voltage or current that can be negative or positive, produced at the output of a device or circuit.

P-channel—A field effect transistor so constructed that p-type semiconductor material is used for the channel through the device.

Panelboard—A metallic or nonmetallic panel used to mount electrical controls, devices or equipment.

Parallel circuit—A circuit that has more than one path for current flow.

Peak-inverse/peak-reverse voltage—The rating of a semiconductor device that indicates the maximum amount of voltage in the reverse direction that can be applied to the device.

Peak-to-peak voltage—The amplitude of voltage measured from the negative peak of an ac waveform to the positive peak.

Peak voltage—The amount of voltage of a waveform measured from the zero voltage point to the positive or negative peak.

Phase shift—A change in the phase relationship between two quantities of voltage or current.

Photodetector—A device that responds to a change in light intensity.

Photodiode—A diode that will conduct in the presence of light and not conduct when in darkness.

Pneumatic timer—A device that uses the displacement of air in a bellows or diaphragm to produce a time delay.

Polarity—The characteristic of a device that exhibits opposite quantities within itself: positive and negative.

Potentiometer—A variable resistor with a sliding contact that is used as a voltage divider.

Power rating—The rating of a device that indicates the amount of current flow and voltage drop that can be permitted.

Pressure switch—A device that senses the presence or absence and causes a set of contacts to open or close.

Pulse generator—An oscillator that produces a voltage of short duration on regular intervals.

RC time constant—The time constant of a resistor and capacitor connected in series. The time in seconds is equal to the resistance in ohms multiplied by the capacitance in farads.

Reactance—The opposition to current flow in an ac circuit offered by pure inductance or pure capacitance.

Rectifier—A device or circuit used to change ac voltage into dc voltage.

Regulator—A device that maintains a quantity at predetermined level.

Relay—A magnetically-operated switch that may have one or more sets of contacts.

Resistance—The opposition to current flow in an ac or dc circuit.

Resistive temperature detector—A device that changes its resistance with a change in temperature. These devices are made of metal and exhibit a positive temperature coefficient.

Ripple—The ac component in the output of a dc power supply caused by improper filtering.

RMS value—The value of ac voltage that will produce as much power when connected across a resistor as a like amount of dc voltage.

Root-mean-square value—*See* rms value.

RTD—*See* resistive temperature detector.

Saturation—The maximum amount of magnetic flux a material can hold.

Schematic—An electrical diagram that shows components in their electrical sequence without regard for physical location.

SCR (silicon-controlled rectifier)—A four-layer semiconductor device that is a rectifier and must be triggered by a pulse applied to the gate before it will conduct.

Semiconductor—A material that contains four valence electrons and is used in the production of solid-state devices. The most common semiconductors are silicon and germanium.

Series aiding—Two or more voltage-producing devices connected in series in such a manner that their voltages add to produce a higher total voltage.

Series circuit—An electric circuit formed by the connection of one or more components in such a manner that there is only one path for current flow.

Signal generator—A test instrument used to produce a low value ac voltage for the purpose of testing or calibrating electronic equipment.

Silicon-controlled rectifier—*See* SCR.

Sine-wave voltage—A voltage waveform whose value at any point is proportional to the trigonometric sine of the angle of the generator producing it.

Solenoid—A magnetic device used to convert electrical energy into linear motion.

Solenoid valve—A valve operated by an electric solenoid. *See* solenoid.

Solid-state device—An electronic component constructed from semiconductor material.

Stealer transistor—A transistor used in such a manner as to force some other component to remain in the off state by shutting its current to electrical ground.

Step-down transformer—A transformer that produces a lower voltage at its secondary than is applied to its primary.

Step-up transformer—A transformer that produces a higher voltage at its secondary than is applied to its primary.

Switch—A mechanical device used to connect or disconnect a component or circuit.

Synchronous speed—The speed of the rotating magnetic field of an ac induction motor.

Temperature coefficient—A ratio of the amount of change a temperature sensing device makes when compared to the amount of change in temperature.

Thermal compound—A grease-like substance used to thermally bond two surfaces together for the purpose of increasing the rate of heat transfer from one object to another.

Thermistor—A device that changes its resistance with a change in temperature. These devices are made of metal oxides and exhibit a negative temperature coefficient.

Thyristor—An electronic component which has only two states of operation; on or off.

Transistor—A solid-state device made by combining three layers of semiconductor material together. A small amount of current flow through the base-emitter can control a larger amount of current flow through the collector-emitter.

Triac—A bidirectional thyristor device used to control ac voltage.

Trigger pulse—A voltage or current of short duration used to activate the gate, base, or input of some electronic device.

Truth table—A chart used to show the output condition of a logic gate or circuit as compared to different conditions of input.

UJT (unijunction transistor)—A special transistor that is a member of the thyristor family of devices and operates like a voltage controlled switch.

Unijunction transistor—*See* UJT.

Valence electron—The electron in the outermost shell or orbit of an atom.

Variable resistor—A resistor whose resistance value can be adjusted between the limits of its minimum and maximum value.

Volt/voltage—An electrical measure of potential difference, electromotive force, or electrical pressure.

Voltage divider—A series connection of resistors used to produce different values of voltage drop across them.

Voltage drop—The amount of voltage required to cause an amount of current to flow through a certain value of resistance or reactance.

Voltage follower—A method of connection for an operational amplifier that will produce a gain of "1" but causes a great increase in input impedance.

Voltage rating—A rating which indicates the amount of voltage that can safely be connected to a device.

Voltage regulator—A device or circuit that maintains a constant value of voltage.

Voltmeter—An instrument used to measure a level of voltage.

Volt-ohm-milliammeter (VOM)—A test instrument so designed that it can be used to measure voltage, resistance, or milliamperes.

Watt—A measure of true power.

Waveform—The shape of a wave as obtained by plotting a graph with respect to voltage and time.

Wye connection—A connection of three components made in such a manner that one end of each component is connected. This connection is generally used to connect devices to a three-phase power system.

Zener diode—A special diode that exhibits a constant voltage drop when connected in such a manner that current flows through it in the reverse direction.

Zener region—The region current enters into when it flows through a diode in the reverse direction.

Zero switching—A feature of some solid-state relays such that current will continue to flow through the device until the ac waveform returns to zero.

Index